**OXSTALLS
LEARNING CENTRE**
Oxstalls Lane, Gloucester
Gloucestershire GL2 9HW

UNIVERSITY OF
GLOUCESTERSHIRE

WEEK LOAN

WEEK LOAN

1 6 JAN 2004

2 6 MAR 2004

1 3 MAY 2004

·1 4 DEC 2004

1 5 APR 2005

2 0 JAN 2006

2 2 MAY 2006

0 5 DEC 2006

1 3 NOV 2007

1 8 JAN 2010

OVERLOAD, PERFORMANCE INCOMPETENCE, AND REGENERATION IN SPORT

OVERLOAD, PERFORMANCE INCOMPETENCE, AND REGENERATION IN SPORT

Edited by

Manfred Lehmann
Medical University Hospital Ulm
Ulm, Germany

Carl Foster
University of Wisconsin–La Cr
La Crosse, Wisconsin

Uwe Gastmann
Medical University Hospital Ul
Ulm, Germany

Hans Keizer
University of Limburg Maastricht
Maastricht, The Netherlands

and

Jürgen M. Steinacker
Medical University Hospital Ulm
Ulm, Germany

Kluwer Academic / Plenum Publishers
New York, Boston, Dordrecht, London, Moscow

Library of Congress Cataloging-in-Publication Data

Overload, performance incompetence, and regeneration in sport / edited
by Manfred Lehmann ... [et al.].
    p. ; cm.
    This volume summarizes the proceedings of a workshop held in
November of 1997.
    Includes bibliographical references and index.
    ISBN 0-306-46106-4
    1. Sports--Physiological aspects Congresses.   2. Physical
education and training Congresses.   3. Fatigue Congresses.
4. Stress (Physiology) Congresses.   I. Lehmann, Manfred.
RC1235.093   1999
617.1'027--dc21   99-15839
                                                                  CIP

Proceedings of the International Conference on Overload, Fatigue, Performance Incompetence, and Regeneration, held November 8, 1997, in Ulm, Germany

ISBN 0-306-46106-4

©1999 Kluwer Academic/Plenum Publishers
233 Spring Street, New York, N.Y. 10013

10 9 8 7 6 5 4 3 2 1

A C.I.P. record for this book is available from the Library of Congress

Printed in the United States of America

PREFACE

This volume summarizes the proceedings of the Reisensburg workshop which took place at Reisensburg Castle in November 1997[a]. The castle is built on the site of an ancient Roman compound and situated in the south of Germany at the Danube river. Scientists from Australia, Austria, Belgium, Estonia, Germany, Italy, Netherlands, South Africa, Switzerland, and the United States participated in the workshop. Like the 1996 workshop, the proceedings of which will be published in *Medicine and Science in Sports and Exercise* in 1998, the 1997 workshop also focused on the topic of overtraining in its widest sense to deepen our knowledge in this particularly sensitive field of sports science and sports practice. The authors see the present volume in a context with the proceedings presented by Guten (ed.) "Running Injuries"; Saunders, Philadelphia (1997) and Kreider, Fry, and O'Toole (eds.) "Overtraining in Sport"; Human Kinetics, Champaign IL (1997).

Overtraining, that is, too much stress combined with too little time for regeneration, can be seen as a crucial and threatening problem within the modern athletic community, of which significance can already be recognized reading daily newspapers: "...During the 1996 European championships, a gymnast shook his head almost imperceptibly, closed his eyes briefly and left the arena without looking up. He was fatigue personified. 'Suddenly, I just couldn't do any more. I just wanted to rest'". A look at his schedule showed why. Two international championships in March, world championships in April, European championships in mid-May, Olympic selection trials at the end of May, national championships and additional selection trials in June..." (Süddeutsche Zeitung, 13 May 1996)." A similar clear message is included when a tennis professional says after an ATP final: "...My recent successes are due to less tennis, more regeneration, and the enforced break (due to injuries). I am less exhausted and burnt out than the other players..." (Süddeutsche Zeitung, 27 October 1995). An identical message is mediated when a professional cyclist states after his success in the Giro d'Italia 1995: "...I am better this year because I trained less. In other years, I was already tired before the race..." (Neue Zürcher Zeitung, 28 May 1995), and when a former soccer professional notes: "...I always ask myself why the others don't learn from us. It is not necessary to train so much to be good..." (Süddeutsche Zeitung, 2 September 1995), or when another tennis professional states after losing the opening match in Indian Wells: "...I have to take a break, otherwise, I can't keep up the pace..." (Stuttgarter Zeitung, 14 March 1996), and the leading champion in the women's skiing world cup notes: "...It's unbelievable, I am surprised myself, the less I train, the better I race..." (Badische Zeitung, 29 December 1997). A quite similar message was expressed by a tennis professional who noted: "...End of 1995 I had to play in the Davis Cup and was tired when I got to Melbourne (beaten in the second round). This time I had a real rest period. I didn't touch the racket for two weeks. I am fresh and ready to go..." (Süddeutsche Zeitung, 9 January 1997), and he became the 1997 Australian Open champion. Contrary to the leading and favored player in the Women's Master's Tennis Cup, New York 1997 who stated: "...Of course, I also wanted to win here, but you get tired at the end of the year, and if you're not completely with it you lose, that's all (beaten by a player who had been sick for 2 months). Maybe I played too often toward the end of the year, but I can learn from my mistake..." (Badische Zeitung, 22 November 1997). Finally, when the captain of a professional German soccer team stated: "...The team is lacking that 100% mental freshness after the second loss within a week. The team seemed burnt out as a result of their tight schedule..." (Frankfurter Allgemeine Zeitung, 8 December 1997), we have to ask how can athletes avoid and escape that national, European, or international web of different interests surrounding them, particularly in professional sports? We hope this workshop will contribute some answers to these questions.

ACKNOWLEDGMENT

Supported by Deutsche Forschungsgemeinschaft, D-53175 Bonn (4853/114/97) and by Bundesinstitut für Sportwissenschaft, D-50877 Bonn.

Manfred Lehmann
Carl Foster
Uwe Gastmann
Hans Keizer
Jürgen Steinacker

CONTENTS

DEFINITION, TYPES, SYMPTOMS, FINDINGS, UNDERLYING MECHANISMS, AND FREQUENCY OF OVERTRAINING AND OVERTRAINING SYNDROME

Manfred Lehmann, [1] Carl Foster,[2] Uwe Gastmann, [1] Hans Keizer, [3] and Jürgen M. Steinacker[1]

[1]University Medical Hospital Ulm
Department of Sports and Rehabilitation Medicine
Steinhoevelstrasse 9, D-89075 Ulm, Germany
[2]University of Wisconsin-La Crosse
132 Mitchell Hall
La Crosse, Wisconsin 54601
[2]University of Limburg Maastricht
Faculty of Health Sciences, Department of Movement Sciences
P.O.Box 616, 6200 MD Maastricht, The Netherlands

1. DEFINTION OF OVERTRAINING AND OVERTRAINING SYNDROME

From an operational standpoint, overtraining can be defined as stress > recovery (regeneration) imbalance, that is, too much stress combined with too little time for regeneration (7, 15, 20-22). In this context, stress summarizes all individual training, non-training, and competition-dependent stress factors (4, 7, 9–11, 15, 18–22, 38, 39, 44). Particularly, additional exogenous non-training stress factors, such as social, educational, occupational, economic, nutritional factors, travel, and endogenous factors (genetic predisposition) exacerbate the risk of a resulting overtraining syndrome in a completely individual manner (7, 15, 18–22). The term overtraining syndrome describes an impaired state of health which is caused by overtraining and characterized by particular findings (7, 15, 18–22).

Overload, Performance Incompetence, and Regeneration in Sport, edited by Lehmann *et al.*
Kluwer Academic / Plenum Publishers, New York, 1999.

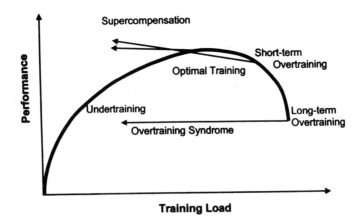

Figure 1. One-dimensional schematic training load–performance relation. Performance improves as training load is increased up to the range of optimal training. At heavier training loads, performance does not show any further improvement, nor deteriorates during a period of overreaching (short-term overtraining). The best performances (supercompensation) are probably obtained with reduced training after mild overreaching (short-term overtraining). If overreaching is too profound or continued too long, performance deteriorates and fails to improve again with reduction in training load. An overtraining syndrome may result. Arrows to the left indicate decrease in training load (regeneration) with (after short-term overtraining) or without improvement in performance (after long-term overtraining). Adapted from Foster C, Lehmann M (1997) Overtraining Syndrome. In:Guten N (ed.) Running Injuries. Saunders, Philadelphia:173–188.

2. TYPES OF OVERTRAINING AND OVERTRAINING SYNDROME

From a prognostic standpoint, short-term overtraining, in general lasting less than 3 weeks, has to be distinguished from long-term overtraining of at least 3 weeks and more (4, 7, 9–11, 15, 20-25, 34, 39, 44), (Figure 1). Short-term overtraining can be seen as a quite usual part of athletic training which, also called overreaching or supercompensation training, leads to a state of overreaching in affected athletes (7, 8–10, 15, 18–22). This state of overreaching is characterized by a transient performance incompetence which is reversible within a short-term recovery period of 1–2 weeks and can be rewarded by a state of supercompensation (16). Supercompensation describes subsequent to 1–2 regeneration weeks a more than expected increase in performance ability after a short-term overtraining-dependent transient period of performance incompetence. The target supercompensation is the reason short-term overtraining or overreaching is accepted as a regular part of athletic training (Figure 1).

However, if overreaching is too profound or extended for too long a time period, in other words, if a necessary regeneration period is inappropriately short and recovery therefore remains incomplete and is additionally associated with too many competitions and non-training stress factors, the athlete clearly runs the risk of a resulting overtraining syndrome. In a simplified training–recovery–model, the transition range between short- and long-term overtraining may be crossed in a particular endurance sport such as running, cycling, swimming, or rowing after (i) 3 weeks of training at a high and increasing training load level of (ii) approximately 3 hours per day, (iii) of more than 30 % increase in training load each week, ignoring (iv) the training principle of alternating hard and easy train-

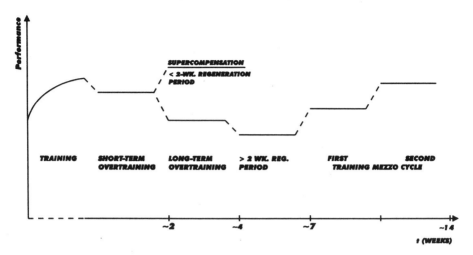

Figure 2. After a short-term overtraining period (overreaching), an improvement in performance (supercompensation) can be expected subsequent to a short-term regeneration period. If overreaching is continued too long or regeneration period was too short, a further decrease in performance cannot be compensated even during an extended regeneration period. An extended regeneration period additionally includes the risk of a further loss in performance so that 1–2 additional training mezzo cycles may be necessary for readaption to an identical performance level as before the overtraining.

ing days, or (v) of alternating 2 hard training days followed by an easy training day, respectively, and not maintaining (vi) a more or less one day training ban per week (21–24). The resulting impaired health state is also called staleness or burnout syndrome (6, 35, 40). Normalization takes an extended long-term recovery period of 2–4 and more weeks (7, 14, 18–22, 44). Such an extended recovery period including an average 30 to 70 % decrease in usual training load, carries the risk of an additional loss in already-reached training adaptation in an affected athlete, therefore, includes the risk of losing the current competition season (Figure 2).

Burnout syndrome or staleness has also been called parasympathetic or Addison type (M. Addison) overtraining syndrome since it resembles a state which is characterized by a predominance in vagal tone or an adrenal insufficiency (14). The more frequent parasympathetic type overtraining syndrome (14, 23–26) has to be distinguished from a less frequent so-called sympathetic or Basedow type (M. Basedow) overtraining syndrome which resembles a hyperadrenergic state or thyreoidal hyperfunction (14).

3. SYMPTOMS AND FINDINGS DURING SHORT- AND LONG-TERM OVERTRAINING

Overtraining key findings are performance incompetence, alteration in mood state, persistent high fatigue ratings, depressed reproductive function, and alteration in immune function (4, 6, 7, 9, 10, 12, 14, 18–22, 28, 30–32, 33, 37, 39, 40, 42), the latter as explained by the so-called open-window theory (30, 33). The definitive key finding of overtraining during periods of intensified load is a lack in progression or a decrease in performance ability which most often is not more than a 0.5–2 % difference. In most sports, however, that is the crucial difference between the champion and the also-ran (7). During short-term overtraining (overreaching) these findings and individually different sports-specific overload complaints are transiently present, less pronounced, and normalized within 2 weeks. After long-term overtraining, however, they persist for more than 2–3 weeks, depending on the duration of the overtraining period.

4. UNDERLYING MECHANISMS

As consequences of the training–competition > recovery imbalance and individually different non-training stress factors, (i) glycogen deficit (5), (ii) catabolic >> anabolic metabolic imbalance (1), (iii) neuroendocrine imbalance (2), (iv) amino acid imbalance (36, 37), and / or (v) autonomic nervous system imbalance (15) are hypothesized as the causes of the above-quoted findings in athletes suffering from an overtraining syndrome. From the standpoint of the overloaded target organs, an overtraining syndrome can be understood as protection against overload-dependent irreversible cellular damage and in part as ultimate negative feedback regulation of the organism (7). This includes (i) decrease in neuromuscular excitability (21–23, 38), (ii) decrease in sensitivity of adrenals to adrenocorticotropic hormone (decreased cortisol release; depressed metabolic competence), (2, 21–23, 28), (iii) decrease in ß-adrenoreceptor density (depressed metabolic and chronotropic competence), (17, 21–23), (iv) decrease in sympathetic intrinsic activity (depressed motivation, drive), (21–25, 27), (v) decreased turnover in contractile proteins (29), (vi) increased synthesis in heat shock proteins (HSP70), (31), (vii) inhibitory effects on first (12) and second alpha-motoneurons (3, 38). The latter, as shown subsequent to acute overload but at present not yet to chronic overtraining, except for a depressed H-reflex (38). Consequences of i, v, vi, vii should be lack in muscular coordination and power. In advanced stages of overtraining, (viii) depressed hypothalamic–pituitary activity can be observed as reflected by depressed responses in adrenocorticotropic and growth hormones thus amplifying the metabolic incompetence (2, 23, 24, 28).

5. FREQUENCY OF OVERTRAINING AND OVERTRAINING SYNDROME

Supercompensation training (short-term overtraining) goes along with an overreaching state in 25–50 % of athletes (23, 24, 43). Findings of long-term overtraining are observed in more than 60 % of distance runners at least once during their career (32), in 20 % of athletes of the Australian swimming team during a season (13, 14), in 33 % of athletes of a national basketball team during a 6-week preparatory period (45), and in more than 50 % of athletes of a soccer team after the first 4-month time period of their competition season (27). Overtraining can thus not be seen as a marginal problem but is an important and frequent event in the athletic community which makes necessary all our efforts to prevent it.

6. FUTURE DIRECTIONS

To deepen international sports-practice and scientific contacts and discussions in this field in order to find ways out of the current misery for our affected athletes, some working groups increased and focused their scientific activities to overreaching, overtraining and regeneration in sports as indicated by the 1996 Memphis Overtraining in Sport Congress (19), the 1996 Reisensburg Castle Overtraining Workshop, of which the proceedings will be published 1998 in a supplement to Medicine and Science in Sports and Exercise, and the 1997 Reisensburg Castle Overtraining Workshop the presentations of which are summarized in this volume.

REFERENCES

1. Adlercreutz H, Harkonen K, Kuoppasalmi K, Naveri H, Huthamieni H, Tikkanen H, Remes K, Dessipris A, Karvonen J (1986) Effect of training on plasma anabolic and catabolic steroid hormones and their response during physical exercise. Int J Sports Med (Suppl):27–28
2. Barron JL, Noakes TD, Lewy W, Smith C, Millar RP (1985) Hypothalamic dysfunction in overtrained athletes. J Clin Endocrinol Metabol 60:803–806
3. Baur S, Miller R, Liu Y, Freiwald J, Konrad P, Steinacker JM, Lehmann M (1997) Changes in EMG pattern during an hour exhausting cycling. In:Dickhuth HH, Küsswetter W (eds) 35. Deutscher Sportärztekongress Tübingen. Novartis Pharma Verlag Wehr:265
4. Budgett R (1990) Overtraining syndrome. Br J Sports Med 24:231–236
5. Costill DL, Flynn MG, Kirwan JP, Houmard JA, Mitchell JB, Thomas R, Sung HP (1988) Effects of repeated days of intensified training on muscle glycogen and swimming performance. Med Sci Sports Exerc 20:249–254
6. Counsilman JE (1955) Fatigue and staleness. Athletic J 15:16–20
7. Foster C, Lehmann M (1996) Overtraining syndrome. In: Guten GN (ed) Running injuries. Saunders Philadelphia:173–188
8. Fry AC (1997) The role of training intensity in resistance exercise overtraining and overreaching. In: Kreider RB, Fry AC, O'Toole ML (eds) Overtraining in sport. Human Kinetics Champaign Il USA:107–127
9. Fry RW, Morton AR, Keast D (1991) Overtraining in athletes. An update. Sports Med 12:32–65
10. Fry RW, Morton AR, Keast D (1992) Periodisation and the prevention of overtraining. Can J Sports Sci 17:241–248
11. Hackney AC, Pearman III SN, Nowacki JM (1990) Physiological profiles of overtrained and stale athletes: a review. Appl Sport Psychology 2:21–33
12. Höllge J, Kunkel M, Ziemann U, Tergau F, Geese R, Reimers CD (1997) Central fatigue during exercise. A magnetic stimulation study. In:Dickhuth HH, Küsswetter W (eds) 35. Deutscher Sportärztekongress. Novartis Pharma Verlag Wehr:209
13. Hooper SL, Mackinnon LT, Gordon RD, Bachmann AW (1993) Hormonal responses of elite swimmers to overtraining. Med Sci Sports Exerc 25:741–747
14. Hooper SL, Mackinnon LT, Howard A, Gordon RD, Bachmann AW (1995) Markers for monitoring overtraining and recovery. Med Sci Sports Exerc 27:106–112
15. Israel S (1976) Zur Problematik des Übertrainings aus internistischer und leistungsphysiologischer Sicht. Medizin und Sport 16:1–12
16. Jeukendrup AE, Hesselink MKC, Snyder AC, Kuipers H, Keizer HA (1992) Physiological changes in male competitive cyclists after two weeks ofintensified training. Int J Sports Med 13:534–541
17. Jost J, Weiss M, Weicker H (1989) Unterschiedliche Regulation des adrenergen Rezeptorsystems in verschiedenen Trainingsphasen von Schwimmern und Langstreckenläufern. In: Böning D, Braumann KM, Busse MW, Maassen N, Schmidt W (Hrsg) Sport, Rettung oder Risiko für die Gesundheit. Deutscher Ärzteverlag Köln:141–145
18. Kreider RB (1997) Central fatigue hypothesis and overtraining. In: Kreider RB, Fry AC, O'Toole ML (eds) Overtraining in sport. Human Kinetics Champaign Il USA:309–331
19. Kreider RB, Fry AC, O'Toole ML (1997) Overtraining in sport. Human Kinetics Champaign Il USA
20. Kuipers H, Keizer HA (1988) Overtraining in elite athletes. Sports Med 6:79–92
21. Lehmann M, Foster C, Keul J (1993) Overtraining in endurance athletes. A breif review. Med Sci Sports Exerc 25:854–862
22. Lehmann M, Gastmann U, Steinacker JM, Heinz N, Brouns F (1995) Overtraining in endurance sports. A short overview. Med Sport Boh Slov 4:1–6

23. Lehmann M, Foster C, Netzer N, Lormes W, Steinacker JM, Liu Y, Opitz-Gress A, Gastmann U (1997) Physiological responses to short- and long-term overtraining in endurance athletes. In: Kreider RB, Fry AC, O'Toole ML (eds) Overtraining in sport. Human Kinetics Champaign Il USA:19–46

24. Lehmann M, Foster C, Steinacker JM, Lormes W, Opitz-Gress A, Keul J, Gastmann U (1997) Training and overtraining: Overview and experimental results. J Sport Med Phys Fitness 37:7–17

25. Lehmann M, Gastmann U, Petersen KG, Bachl N, Seidel A, Khalaf AN, Fischer S, Keul J (1992) Training–overtraining: performance and hormone levels, after a defined increase in training volume vs. intensity in experienced middle- and long-distance runners. Br J Sports Med 26:233–242

26. Lehmann M, Baumgartl P, Wiesenack C, Seidel A, Baumann H, Fischer S, Spöri U, Gendrisch G, Kaminski R, Keul J (1992) Training–overtraining: influence of a defined increase in training volume vs. training intensity on performance, catecholamines and some metabolic parameters in experienced middle- and long-distance runners. Eur J Appl Physiol 64:169–177

27. Lehmann M, Schnee W, Scheu R, Stockhausen W, Bachl N (1992) Decreased nocturnal catecholamine excretion: parameter for an overtraining syndrome in athletes ? Int J Sports Med 13:236–242

28. Lehmann M, Knizia K, Gastmann U, Petersen KG, Khalaf AN, Bauer S, Kerp L, Keul J (1993) Influence of 6-week, 6 days per week, training on pituitary function in recreational athletes. Br J Sports Med 27:186–192

29. Liu Y, Opitz-Gress A, Steinacker JM, Zeller C, Mayer S, Lormes W, Gastmann U, Altenburg D, Lehmann M (1997) Adaptation of muscle fibers to 4-week competition preparing training in rowers. In:Dickhuth HH, Küsswetter W (eds) 36. Deutscher Sportärztekongress Tübingen. Novartis Pharma Verlag Wehr:162

30. Mackinnon LT (1997) Effects of overreaching and overtraining on immune function. In: Kreider RB, Fry AC, O'Toole ML (eds) Overtraining in sport. Human Kinetics Champaign Il USA:219–241

31. Mayr S, Steinacker JM, Opitz-Gress A, Liu Y, Zeller C, Lormes W, Lehmann M (1997) Muscular stress reaction to 4-week competition preparation training in rowers. In:Dickhuth HH, Küsswetter W (eds) Novartis Pharma VerlagWehr:164

32. Morgan WP, Brown DR, Raglin JS, O'Connor PJ, Ellickson KA (1987) Psychological monitoring of overtraining and staleness. Br J Sports Med 21:107–114

33. Nieman DC (1997) Effects of athletic endurance training on infection rates and immunity. In: Kreider RB, Fry AC, O'Toole ML (eds) Overtraining in sport. Human Kinetics Champaign Il USA:193–217

34. O'Toole ML (1997) Overreaching and overtraining in endurance athletes. In: Kreider RB, Fry AC, O'Toole ML (eds) Overtraining in sport. Human Kinetics Champaign Il USA:3–17

35. Owen IR (1964) Staleness. Phys Educ 56:35

36. Parry-Billings M, Blomstrand E, McAndrew N, Newsholme N, Newsholme EA (1980) A communicational link between skeletal muscle, brain, and cells of the immune system. Int J Sports Med 11 (Suppl 2):122–128

37. Parry-Billings M, Budgett R, Koutedakis Y, Blomstrand E, Brooks S, Williams C, Calder PC, Pilling S, Baigrie R, Newsholme EA (1992) Plasma amino acid concentration in the overtraining syndrome: possible effects on the immune system. Med Sci Sports Exerc 24:1353–1358

38. Raglin JS, Kocera DM, Stager JM, Harms CA (1996) Mood, neuromuscular function, and performance during training in female swimmers. Med Sci Sports Exerc 28:372–377

39. Rowbottom DG, Keast D, Morton AR (1997) Monitoring and preventing of overreaching and overtraining in endurance athletes. In: Kreider RB, Fry AC, O'Toole ML (eds) Overtraining in sport. Human Kinetics Champaign Il USA:47–66

40. Rowland TW (1986) Exercise fatigue in adolescents: diagnosis of athlete burnout. Phys Sportsmed 14:69–77

41. Snyder AC, Jeukendrup AE, Hesselink MKC, Kuipers H, Foster C (1993) A physiological / psychological indicator of overreaching during intensive training. Int J Sports Med 14:29–32

42. Snyder AC, Kuipers H, Chang BO, Servais R, Fransen E (1995) Overtraining following intensified training with normal muscle glycogen. Med Sci Sports Exerc 27:1063–1070

43. Steinacker JM, Lormes W (1998) Findings during overload and regeneration in elite rowers between national and world championships: This volume

44. Stone MH, Keith RE, Kearney JT, Fleck SJ, Wilsond GD, Triplett NT (1991) Overtraining. A review of signs, symptoms and possible causes. J Appl Sport Science Research 5:35–50

45. Verma SK, Mahindroo SR, Kansal DK (1978) Effect of four weeks of hard physical training on certain physiological and morphological parameters of basket-ball players. J Sports Med 18:379–384

2

SELECTED PARAMETERS AND MECHANISMS OF PERIPHERAL AND CENTRAL FATIGUE AND REGENERATION IN OVERTRAINED ATHLETES

Manfred Lehmann, Uwe Gastmann, Susanne Baur, Yufei Liu, Werner Lormes, Alexandra Opitz-Gress, Susanne Reißnecker, Christoph Simsch, and Jürgen M. Steinacker

Department of Sports and Rehabilitation Medicine
University Medical Hospital Ulm
Steinhoevelstrasse 9
D-89073 Ulm, Germany

1. INTRODUCTION

Definition, types, symptoms, findings, underlying mechanisms, and frequency of overtraining and overtraining syndrome have been described in an introductory article to the present volume (51). During the past 10 years, our increasing knowledge in this field has also been discussed in different original and review articles (6, 8, 11, 12, 13, 14, 15, 23, 30, 35, 37, 41, 42, 44, 52, 53, 55, 57, 60–62, 68, 69, 75, 76, 78, 80, 81), as well as presented at the 1996 Memphis Overtraining and Overreaching in Sports Conference and summarized in a book project (36). Aim of this present overview, which was presented during the 1997 Reisensburg Castle workshop, is an additional up-dating of our knowledge considering mechanisms underlying overtraining-related performance incompetence in affected athletes with respect to further results obtained in this field during the past 2 years. Particular emphasis has been given to the time-course of regeneration subsequent to overtraining as far as it is known at present. From an operational standpoint, the thesis was followed that findings such as impairment of neuromuscular function (47, 48, 67), depressed ß-adrenergic receptor density (32), related depressed lipolysis, glycogenolysis, glycolysis, and heart rate response (43, 44), as well as depressed intrinsic sympathetic activity (42–44, 49, 58, 77), depressed turnover in contractile proteins (54, 72–74), depressed adrenocortical (2, 51, 80), and pituitary–hypothalamic responsiveness in an advanced stage (2), or iron deficiency (53) can explain performance incompetence in overtrained athletes, whereas appropriate regeneration should be indicated by their normalization.

Overload, Performance Incompetence, and Regeneration in Sport, edited by Lehmann *et al.*
Kluwer Academic / Plenum Publishers, New York, 1999.

2. SELECTED PERIPHERAL MECHANISMS UNDERLYING PERFORMANCE INCOMPETENCE AND THEIR NORMALIZATION IN OVERTRAINED ATHLETES

Peripheral mechanisms, i.e., mechanisms not originating in the brain are summarized in Figure 1 which also includes selected central mechanisms. There is now convincing evidence that these mechanisms are involved in overtraining-related performance incompetence.

2.1. Neuromuscular Excitability (NME)

NME can be measured using the minimal current pulse at different impulse durations which is necessary to induce a single contraction of a respective skeletal muscle fiber (65). NME is improved in trained reference muscles of endurance-trained athletes (4). Depressed NME of fatigued muscles is already found after a single bout of prolonged heavy exercise and shows normalization after neight rest (4, 71). Since in a one-leg exhausting exercise test, this finding can only be demonstrated in muscles of the stressed leg, this fatigue reaction primarily affects peripheral neuromuscular structures rather than central mechanisms (71). Prospective research also revealed a more profound decrease in

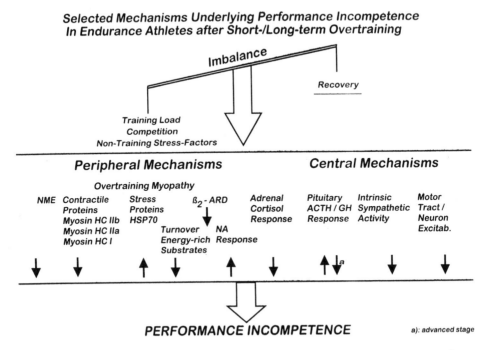

Figure 1. Selected peripheral and central mechanisms involved in overtraining-related performance incompetence. Neuromuscular excitability (NME) turnover in contractile proteins, $\beta_2$adrenoreceptor density (β_2ARD), related turnover in energy-rich substrates, and adrenocortical cortisol response are depressed in an early stage, whereas neuronal noradrenaline response (NA), pituitary ACTH and GH responses, as well as shock protein synthesis (HSP 70) are increased. In an advanced stage, pituitary ACTH, pituitary GH responses, and intrinsic sympathetic activity are depressed. Additionally, there is increasing evidence of depressed motor tract / motor neuron excitability in fatigued athletes (25, 47, 48, 67).

NME of respective reference muscles subsequent to overtraining at high-energetic demands (54) which was not normalized after night rest. This finding was confirmed in cyclists even after overtraining at moderate-energetic demands (48). After a 2-week regeneration period, impaired NME was again normalized (Figure 2).

Since there was no additional monitoring of NME during these 2 regeneration weeks, we can not say how many days are definitively necessary for complete normalization of NME subsequent to a 3- or 6-week period of intensified training (48). There is, however, additional evidence in elite rowers that normalization of NME can already be observed after only 1 week of regeneration subsequent to a 3-week period of intensified training (Baur, S.: Still unpublished results). Since, however, normalization of NME after a 1–2 week regeneration period

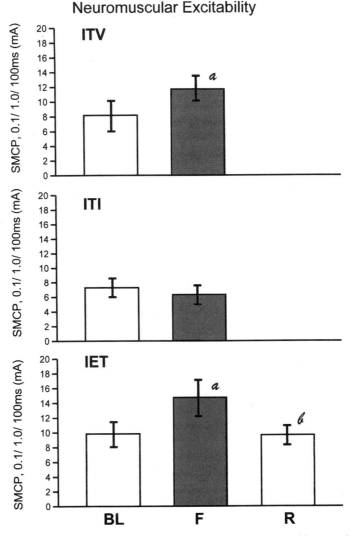

Figure 2. At final examination (F) compared to baseline (BL), NME was depressed in overtrained distance runners after the 4-week increase in training volume study (ITV), (45–47). This was also oberserved in cyclists after the 6-week intensified ergometer overtraining study (IET), (48), as compared to the 4-week increase in training intensity control study (ITI), (47). Depressed NME was again normalized after 2 regeneration weeks (R) as observed in the cyclists (IET). SMCP: Summed minimal current pulse at different impulse durations which is necessary to induce a single contraction of vastus medialis muscle fibers.

did not coincide with normalization of performance ability in the overtrained cyclists (48) and the overtrained elite rowers (Baur, S.: Still unpublished results), normalization of NME does not necessarily indicate adequate regeneration with respect to normalization of performance ability. Therefore, examination of NME can be used to monitor training in order to detect overtraining, but not to monitor regeneration to detect normalization of performance ability. Accelerated catabolic cortisol effects on sarcolemma proteins are likely a significant factor for deterioration in the excitability of skeletal muscle sarcolemma or neuromuscular excitability in glucocorticoid supplemented and/or exercising laboratory animals (19–22, 33, 34, 72–74). There is evidence of an identical mechanism in human athletes, since during prolonged heavy exercise, deterioration of NME was prevented by carbohydrate supplementation which caused a 30% decrease in ACTH and free cortisol responses, in contrast to the control experiment (16, 17), (Figure 3).

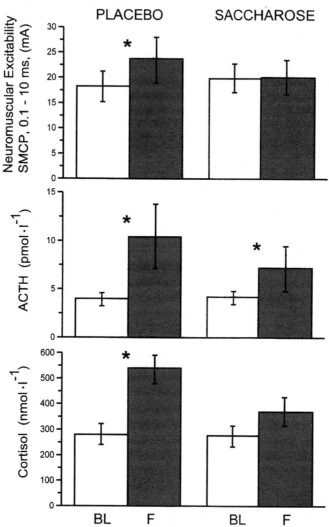

Figure 3. NME was depressed in cyclists after a 2-hour exhausting exercise period (PLACEBO), (16, 17). This went along with excess serum / plasma ACTH and cortisol concentrations. Excess cortisol concentrations and depressed NME were prevented by carbohydrate supplementation (SACCHAROSE). SMCP, BL, F: see Fig. 2; *: significant compared to BL.

2.2. Contractile Myofibrillar and Heat Shock Proteins

The known catabolic effect of glucocorticoids on sarcolemma proteins as cited above, and on skeletal muscle contractile proteins (19–22, 33, 34, 72–74) is seen to be realized through augmented alkaline myofibrillar activity dependent on increased mast cell activity and their degranulation of alkaline proteases as well as on increased lysosomal activity (72–74). At present, the mechanism by which alkaline proteases enter muscle cells is unknown, as is the reason why glucocorticoids do not affect protein synthesis in heart muscle fibers and significantly less in oxidative skeletal muscle fibers (72–74). Beside decreased incorporation of amino acids into myosin heavy chains (19–22, 74), altogether, there is some evidence of an depressed turnover of myosin heavy chains glycolytic type II >> oxidative type II > oxidative type I fibers or actin (72–74). On the one hand, quite similar excess plasma glucocorticoid concentrations can be observed in Cushing's syndrome (66) and in healthy athletes during prolonged exhaustive exercise (50, 53) (Figure 4).

On the other hand, a pronounced glucocorticoid myopathy is a common finding in Cushing's Syndrome and in glucocorticoid supplemented laboratory animals (72–74). Since this myopathy is quite similar in glucocorticoid supplemented rats and in not-supplemented but overtrained rats after 2 weeks of swimming overtraining (72–74), long-term increased overtraining-related (endogenous) cortisol levels might be an important pathogenetic factor underlying such an overtraining myopathy. Therefore, there is some evidence that a mild form of such a glucocorticoid myopathy can also be expected in overtrained human athletes (54). Beside an impaired NME which was observed in overtrained distance runners (47), cyclists (48), and elite rowers (Baur, S.: Still unpublished results.), this conclusion can be drawn from findings in the overtrained elite rowers who

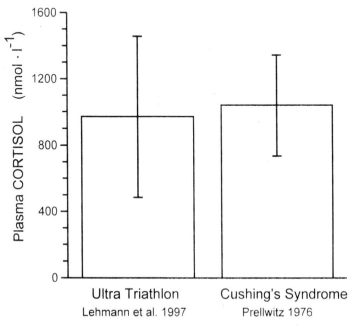

Figure 4. Excess serum cortisol levels can be quite similar in patients suffering from Cushing's syndrome (66) and in athletes during an ultra event (50, 53).

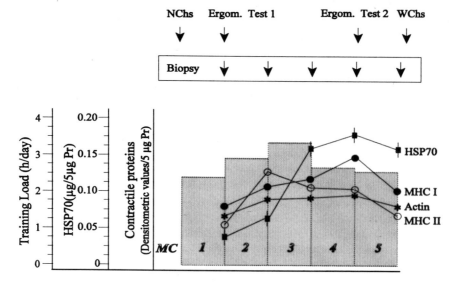

Figure 5. Turnover of contractile proteins in elite rowers during a period of overreaching (MC: micro cycles 1–3) and tapering (MC 4 and 5), (54). After MCs 1 and 2 of intensified training, turnover of myosin heavy chain type II fibers (MHCII), type I fibers (MHCI), and actin was increased compared to baseline. After 2–3 MCs of intensified training; MHCII continously decreased even during the 2-week regeneration or tapering period (MC 4 and 5). Actin turnover remained nearly constant, but MHCI showed a "late" decrease during MC 5 subsequent to World Championships.

simultaneously showed a depressed turnover of myosin-heavy-chains type II fibers, less type I fibers, or actin during a 2–3 week overreaching period (54) which was not normalized after a 2-week regeneration period (Figure 5).

Thus, it may be concluded that muscular fatigue (41–44), muscular soreness (26, 27), and decreased muscular performance ability (10, 13, 26, 27, 31, 45, 46, 69, 75–77, 80) in overtrained athletes can indicate a mild form of "glucocorticoid myopathy" of which we do not know at present how long it takes for complete regeneration. Furthermore, we do not know whether complete normalization of this mild myopathy is necessary for complete normalization of depressed muscular performance ability in overtrained athletes. In our opinion, muscle cell leakage, as indicated by a significant increase in serum activity of muscular enzymes, is not an integral prerequisite of such an overtraining myopathy. This is rather a sign of an exercise myopathy dependent on muscular (micro)-injuries subsequent to acute muscular overload such as during an ultratriathlon (50). Contradictory to the observed decreased turnover in myofibrillar contractile proteins, overtrained athletes show a persistent high myofibrillar synthesis or concentration of 70 kDa heat shock proteins (HSP 70) during a prolonged period of intensified training (56) which did not show complete normalization even after a 2-week regeneration period. These proteins increase stress tolerance of affected cells and conduct cellular repair processes (70). Coincidental to a still suppressed myofibrillar turnover of contractile proteins, a persistent high HSP 70 synthesis in the same athletes may indicate a state of inadequate regeneration even after a 2-week regeneration period subsequent to 2–3 weeks of intensified daily training in elite rowers (44). However, these findings must not necessarily coincide with a still-depressed performance ability subsequent to a 2-week regeneration period, since these affected athletes finished World Championships very successfully (44). However, the number of medals can not be accepted as a sufficient argument for an adequate regeneration. Furthermore, the mechanism is presently not known why chronically increased circulating glucocorticoids impose a catabolic effect on sarcolemma proteins as well as on myofibrillar skeletal muscle proteins but not on stress protein synthesis.

2.3. Beta₂-Adrenoreceptor Density (ßARD)

Beside decreased ßARD which was observed in swimmers (9), the majority of findings indicates higher ßARD on intact blood cells or membrane fractions of blood cells in well-trained and well-regenerated runners (5, 32, 38) and swimmers (32) than in untrained controls. This goes along with higher metabolic and left-ventricular cardiac sensitivity to catecholamines (38, 39). But, this increased sensitivity to catecholamines was not observed for the cardiac sinus node likely because of a simultaneous increase in vagal tone. After a 3- or 5-week period of high-volume endurance training, a decrease in ßARD on intact blood cells (33, 38, 39) and in isoproterenol-stimulated cyclo-AMP activity (32) towards the level of untrained controls were described in elite distance runners and swimmers. This may reflect an overload-dependent loss in adaptation to endurance exercise as compared to recovery or tapering periods (Figure 6).

. Performance ability was not systematically monitored in these athletes in the 80s when these studies took place (32, 38, 39) because overtraining was not yet known to be such an important problem in the athletic community, so there is no information concerning a well-trained or overtrained state in these athletes. However, since a decrease in ßARD already occurred during such high-volume training periods, it should in any case occur after high-volume overtraining periods. In agreement with the assumption of a de-

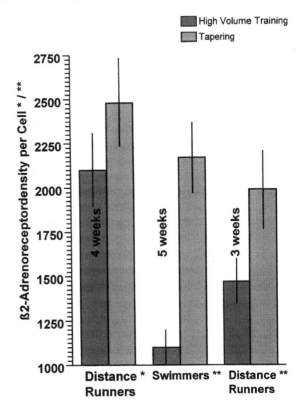

Figure 6. Beta₂-adrenoreceptor density is depressed in distance runners (32, 38, 39) and swimmers (32) after 3–5 weeks high-volume training periods as compared to tapering (*adapted from Lehmann et al., 1983/1984 [polymorphonuce. Leucocytes], **adapted from Jost et al., 1989 [Lymphocytes]).

creased ßARD, overtrained distance runners showed depressed serum glucose, lactic acid, free fatty acid, and heart rate responses combined with increased resting plasma noradrenaline levels (43, 44) and submaximum plasma noradrenaline responses (43, 44) after a 3–4-week high-volume overtraining period. Some of these overtrained distance runners commented on this situation with typical phrases like "You step on the gas and nothing happens!" The pattern of findings (42, 43) was identical to that observed after intake of beta-blockers (40). Increased resting plasma noradrenaline levels were independently confirmed by Hooper et al. (26, 27) in overtrained swimmers, increased submaximum noradrenaline responses by Fry et al. (13) in athletes after resistance overtraining. Therefore, an underlying decrease in ßARD is rather the explanation for the increased neuronal noradrenaline release, the depressed responses in energy-rich substrates and presumably in energy flux rate, as well as in submaximum heart rate response than of underlying reduced stores in energy-rich substrates. This can be assumed since (i) an overtraining state was also observed with normal glycogen stores (75, 76), since (ii) exhausted glycogen stores can not explain depressed heart rate responses, and (iii) a profound exhaustion of triglyceride stores is not imaginable under these conditions as the cause of a depressed free fatty acid response. A persistent depression in ßARD, as well as in related metabolic and heart rate responses can only be expected during daily overtraining at high energetic demands of more than 4000 kcal per day indicating extended daily training periods with an respective overload of receptors and post-receptor mechanisms (45, 46). This was not observed during overtraining at moderate energetic demands of less than 3000 kcal per day (51). The explanation might be that there remained enough time daily for regeneration of receptors. The first step or a short-term mechanism to increase/decrease adrenoreceptors can be understood as an agonist-dependent externalization/internalization of preformed membrane-bound receptors which works within 15 to 30 or 60 minutes and is reversible in similarly short time periods (7, 79). Long-term effects on receptor density may additionally depend on more profound mechanisms such as an increase/decrease in turnover of respective protein synthesis. Additionally, the overtraining-dependent decrease in adrenoreceptor density may depend on the same catabolic cortisol action on sarcolemma proteins through degranulation of mast cell's alkaline proteases, that is, also on proteins of membrane-bound receptors, as described for neuromuscular structures and myofibrillar contractile proteins of skeletal muscle cells (72–74). Now, decisive questions are, (i) how much time does regeneration take after an overtraining period to attain the same adrenoreceptor density level (adaptation level) as before the overtraining period, and, (ii) is that possible only by reducing training load during a respective regeneration time period, or (iii) must the regeneration period be followed by 1 or 2 additional training micro cycles? At present, these questions can not be answered, since there is no prospective study monitoring receptor density during overtraining and a subsequent regeneration period. The only and unsatisfactory answers we can presently give to these questions are based on the 1988 distance runner overtraining study at high energetic demands (45, 46) after which the athletes showed the typical depression in metabolic and heart rate responses (53). These athletes failed to equal or improve their personal records during a subsequent 3-month follow-up, pointing to the necessity of an extended regeneration and readaptation period. This is in agreement with earlier and similar anecdotal observations ("anecdotal", since regeneration/readaptation was not carefully monitored in these athletes) in distance runners (32, 38, 39) and swimmers (32) in whom regeneration of adrenoreceptor density took a tapering period of some weeks. Altogether, we can conclude that adrenoreceptor density significantly coincides with performance ability. Furthermore, there is some evidence that readaptation in this respect takes more than 2 or 3 weeks. But finally we have to consider that there is at present no study during which ßARD on blood cells as well as on skeletal muscle cells including subsequent metabolic effects were simultaneously studied during a period of overtraining and recovery.

2.4. Adrenal Cortisol Release

Depressed adrenal cortisol responsiveness to exogenous adrenocorticotropic hormone was observed in chronically fatigued horses (6) as well as in overtrained cyclists using corticotropin releasing hormone testing (51). In agreement with these findings, decreased resting (45, 75, 76) and exercise-stimulated maximum cortisol responses (80) were observed in overtrained athletes. This was, however, not reflected in overtrained distance runners by a significantly lower 24-h excretion of free cortisol in urine (45, 46). Thus, measuring 24-h urinary cortisol excretion is not sensitive enough to detect small but significant differences in baseline or maximum adrenal responsiveness. During this state, corticotropin releasing hormone-stimulated pituitary adrenocorticotropin release was about 50–70 % higher in overtrained cyclists than at baseline (51) reflecting a positive hypothalamo-pituitary feedback mechanism to depressed adrenal responsiveness (Figure 7).

There is some evidence that the first step in this adrenal adaptation to overtraining-related overload includes depressed adrenal responsiveness to adrenocorticotropic hormone (51, 82) or exercise-related stimulation (80) which has to be interpreted as a protective mechanism against chronic overload. According to an agonist-dependent downregulation of ß-adrenoreceptors subsequent to high-volume training/overtraining (32), a down regulation of adrenal adrenocorticotropic receptors may be the reason for depressed adrenal sensitivity to adrenocorticotropic hormone. In agreement with increased neuronal release of noradrenaline as related to decreased ßARD (13, 26, 27, 45, 46), this is answered by increased pituitary release of adrenocorticotropic hormone in an early state of

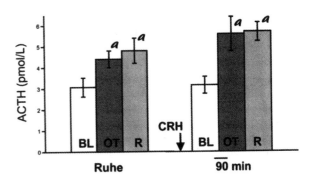

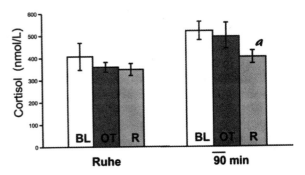

Figure 7. In cyclists, after a 6-week high-intensity overtraining period (OT), pituitary ACTH response to corticotropin releasing hormone (CRH test) was significantly increased compared to baseline (BL). This went along with a slightly decreased adrenocortical cortisol response. This points to a depressed adrenocortical sensitivity to ACTH or to a still unknown inhibitory mechanism. After an additional insufficient 2-week regeneration period (R), pituitary ACTH response was still increased but adrenocortical cortisol response was even more depressed (51).

overtraining or overreaching (51, 82). However, neither increased neuronal noradrenaline release nor increased pituitary release of adrenocorticotropic hormone can overcome a depressed sensitivity of target organs to these hormones. The slightly reduced adrenocortical cortisol release lagged behind and was even amplified after 2 weeks of incomplete recovery (51); "incomplete recovery" on the one hand, since performance incompetence still persisted in these overtrained cyclists after 2 regeneration weeks (51), and on the other hand, since 2 weeks of regeneration are not enough to regenerate adrenal sensitivity to adrenorcorticotropic hormone. In ultra-triathletes, increased release of adrenocorticotropic hormone was also described by Wittert et al. (82) 3–5 days subsequent to the one-day coast-to-coast New Zealand ultra triathlon compared to untrained controls. Since both groups showed similar cortisol levels, the cortisol: ACTH ratio was clearly depressed in the athletes, thus indicating depressed adrenal responsiveness to ACTH in agreement with the above-quoted findings in overtrained cyclists (51). Furthermore, since after such an ultra-event and subsequent to a heavy preparatory season, 3–5 days are too short for complete regeneration, the findings rather reflect an state of overreaching or incomplete recovery in agreement with the results in the overtrained cyclists (51). Gastmann et al. (18) also described a clearly increased pituitary release of adrenocorticotropic hormone in cyclists quite near the end of their competition season, reflecting an overtraining state as a typical late season disease. To summarize, in healthy athletes we can see increased pituitary release of adrenocorticotropic hormone combined with depressed adrenal cortisol release as an indicator of an early overtraining state as observed in overtrained cyclists (51) and incompletely regenerated ultra triathletes (82). Like depressed ß-adrenoreceptor density (32), depressed cortisol release contributes to metabolic incompetence and therefore to performance incompetence in overtrained athletes. Since complete regeneration takes more than 2 weeks, affected athletes run the risk of losing the competition season because an extended regeneration period of more than 2 weeks likely goes along with an additional loss in endurance performance adaptation (28), making 1 or 2 additional training micro cycles necessary to again develop a similar performance ability as before the overtraining.

3. SELECTED CENTRAL MECHANISMS

In this section, selected "central" mechanisms are discussed of which there is some evidence that they have to be understood as negative feedback regulation to an overtraining-related overload of target organs, i.e. to mechanisms originating in the periphery. Therefore, these "central" mechanisms can not be seen as originating exclusively in the brain during an overtraining period but as part of a superior control system. Furthermore, endocrinological data and mechanisms are also discussed in the contributions of A.C. Hackney as well as A. Viru and M. Viru to this volume but in a more extended endocrinological context of the overtraining topic. This offers the reader the attractive opportunity to detect whether there is still some uncertainty in interpretations of findings between these working groups or extensive agreement.

3.1. Pituitary ACTH and Growth Hormone Release

In an early stage of the overtraining process, there is a significantly increased pituitary adrenocroticotropin (ACTH) response and slightly increased pituitary growth hormone (GH) response as observed in cyclists (51) and as already discussed in section 2.4. In contrast to these findings, the pituitary ACTH and GH responses were significantly reduced in overtrained distance runners in an advanced stage of the overtraining process (2). Both in this advanced stage as observed in distance runners (2) and in an early stage of the overtraining process as observed in cyclists (51), adrenal cortisol response was clearly reduced. In the early stage, increased pituitary ACTH- and GH responses have to be seen as positive feedback regulation to depressed adrenal ACTH responsiveness and cortisol release. The increased ACTH release, however, failed to overcome completely the depressed adrenal responsiveness (51, 80). The result can be seen as a depressed metabolic competence in overtrained athletes. Similar findings as observed in an early overtraining stage, i.e. a slightly depressed adrenal cortisol response combined with increased ACTH levels, have been described in ultra triathletes compared to healthy controls (82). These triathletes were examined 3–5 days after the one day coast-to-coast New Zealand ultra triathlon. i.e. in a stage of still incomplete recovery, as already discussed in section 2.4. Wittert et al. (82) agreed with this additional interpretation of their data so we can suppose the athletes were still in a stage of overreaching. In an advanced stage of overtraining, the then depressed pituitary ACTH- and GH responses (2) have to be seen as a particular hypothalamic or central step indicating "central fatigue" or an ultimate negative feedback regulation to chronic overload thus amplifying the metabolic incompetence of affected athletes (Figure 8).

The bad news for an affected athlete is that, in this advanced stage, complete normalization takes an extended recovery period of 2–4 weeks (2). A hypothalamic inhibitory > excitatory imbalance has to be assumed as the underlying mechaintion of this "ultimate negative feedback regulation". Factors which can cause such an imbalance include: (i) Negative hormonal feedback based on long-term increased levels of circulating free cortisol, catecholamines, growth hormone, or insuline like growth factors., (ii) an afferent

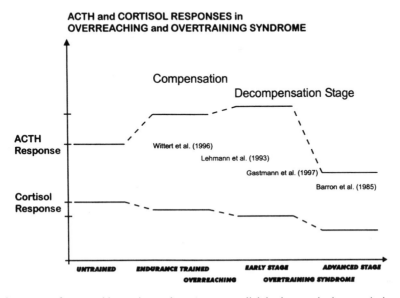

Figure 8. In a stage of overreaching or incomplete recovery, a slightly depressed adrenocortical sensitivity to ACTH is compensated by increased pituitary ACTH release (Compensation Stage), (82). In an early stage of the overtraining syndrome, adrenocortical sensitivity to ACTH is significantly reduced (see Figure 7), (51). In an advanced stage, pituitary ACTH release is also significantly depressed (Decompensation Stage), (2). With permission of Lehmann M, Foster C, Dickhuth HH, Gastmann U (1998) Autonomic imbalance hypothesis and overtraining syndrome. Med Sci Sports Exerc 30:1140–1145.

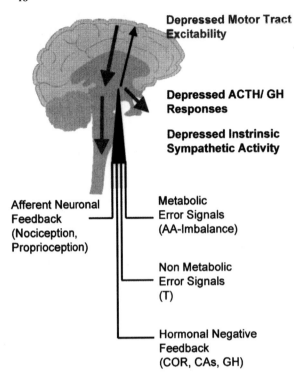

Depressed Motor Tract
Excitability

Depressed ACTH/ GH
Responses

Depressed Instrinsic
Sympathetic Activity

Afferent Neuronal
Feedback
(Nociception,
Proprioception)

Metabolic
Error Signals
(AA-Imbalance)

Non Metabolic
Error Signals
(T)

Hormonal Negative
Feedback
(COR, CAs, GH)

Figure 9. Factors which can cause hypothalamic inhibitory > excitatory imbalance include (i) afferent neuronal negative feedback from overloaded muscles, (ii) hormonal negative feedback based on long-term increased levels of circulating "ergotropic" hormones such as cortisol, catecholamines, or growth hormone, (iii) metabolic error signals as a long-term amino acid imbalance, and (iv) non-metabolic error signals such as long-term increased body core temperature (T: body core temperature; AA: amino acids).

neuronal negative feedback originating in long-term overloaded muscles (nociception, proprioception) using unmyelinated afferent group 3 and 4 fibers, (iii) metabolic error signals (1) such as a long-term amino acid imbalance (35) with an increase in hypothalamic 5-hydroxytryptamin concentration (63, 64), and (iv) non-metabolic error signals such as a long-term increase in body core temperature during extended training periods (Figure 9).

A chronically increased body core temperature can also cause decreased pituitary release of thyroid-stimulating hormone and trijodothyronine as observed in exhausted mountaineers (24). But at present our knowledge concerning the hypothalamic-pituitary-thyroid axis and overtraining is very limited.

3.2. Intrinsic Sympathetic Activity

Nocturnal urinary catecholamine excretion, so-called basal catecholamine excretion, can be seen as an indicator of intrinsic sympathetic activity, since activating mechanisms of the sympathetic nervous system are clearly reduced during night rest (48). Anecdotal observations revealed a significantly (50–70 %) decreased basal catecholamine excretion in overloaded cyclists before the Olympic Games in Seoul (49), as well as in an exhausted top tennis player after playing 3 different ATP finals on 3 different continents within approximately 4 weeks (43, 44). These results have been confirmed during prospective research in overtrained soccer players (49), distance runners (45, 46), (Figure 10), and independently by Naessens et al. (58) in a further group of soccer players. There are negative correlations to fatigue rating (42, 43) and REM sleep latency as observed by Netzer et al. (59) in exhausted cyclists.

The lower basal catecholamine excretion the higher was fatigue rating in overtrained distance runners (42–44) and the shorter REM sleep latency as observed in exhausted cy-

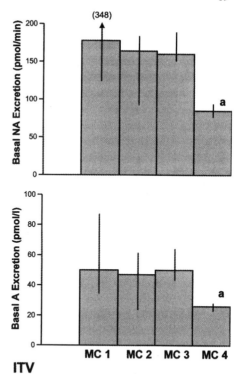

Figure 10. In distance runners, basal adrenaline and no-radrenaline excretions showed a significant decrease during week 4 (micro cycle 4) of a high-volume overtraining study (45, 46), as also observed in overtrained soccer players (49, 58), cyclists (18, 49), and rowers (77).

clists (59). Therefore, basal catecholamine excretion can most likely be related to a complex mechanism causing central fatigue. Basal catecholamine excretion must also be seen in context with motivation and drive. But further research is necessary in this respect. The underlying factors which cause depressed intrinsic sympathetic activity are probably the same as discussed for depressed pituitary responsiveness (Figure 9).

3.3. Depressed Pyramidal Tract and Alpha-Motorneuron Excitability

3.3.1. Depressed Magnetically Evoked Potentials. Excitability of "1st. alpha-motor-neuron" (pyramidal tract) is depressed after exhausting exercise as indicated by a 18–36 % decrease in magnetically evoked potentials after exhaustive anaerobic exercise but this was not observed after moderate aerobic exercise (25). The underlying mechanism can likely be seen in a hypothalamic inhibitory influence via thalamus on pyramidal cells of the brain cortex as discussed in section 3.1. (Figure 9). However, at present, we lack respective findings during a chronic state of fatigue subsequent to overtraining.

3.3.2. Depressed Hoffmann-Reflex (H-Reflex). Raglin et al. (66) observed an impaired neuromuscular function in overtrained swimmers which was measured using the soleus Hoffmann-reflex (H-Reflex). Repeated measures revealed that H-reflex and peak and average swimming power were reduced below baseline whereas total mood disturbance was elevated above baseline values at peak training. These findings returned to baseline during the tapering period. Since so-called M-wave responses were not significantly affected in these overtrained swimmers, Raglin et al. (66) conclude that alteration in H-reflex is the result of a depressed excitability of the alpha-motorneuron rather than a local alteration within the muscle as described in section 2.1.

3.3.3. Selected Electromyographic Findings to Detect Muscular Fatigue. Using electromyography, muscular fatigue is expected to be indicated by a decrease in electromyographic frequency and an increase in integrated and average electromygraphic activity (3). However, this could not be confirmed during exhausting prolonged cycling exercise; since preliminary results revealed a decrease in integrated and average electromyographic activity of vastus medialis and rectus femoris muscles during a standardized exhausting 2-hour cycling period (42). This went along with constant or slightly increased electromyographic frequency. Resently, these findings have been confirmed in part in cyclists during a 1-hour cycling period (3). At the end of this period, compared to baseline, a right-shift, that is a delay in the increase in electromyographic activity was observed for Vastus medialis, Biceps femoris, Tibialis anterior, and Gastrognemius muscles. During the carefully standardized cycling tests, this findings may indicate a delayed muscular activation and probably an impairement in muscular coordination in the state of muscular fatigue. This went along with a decrease in integrated and average electromyographic activity as measured using auxillary muscles that is the smaller Biceps femoris, Tibialis anterior, and Gastrognemius muscles. For these smaller muscle groups it can be expected to show earlier sings of fatigue than larger muscle groups such as Quadriceps muscle (Figures 11 and 12).

If we argue on the one hand that the delayed increase in electromyographic activity may indicate a fatigue-related impairment in coordination, we can not argue on the other hand that a decrease in integrated and average electromyographic activity indicates an improvement in coordination, i.e. an improved recruitment of muscle fibers. Therefore, there is much evidence that both findings reflect muscular and central fatigue in a state of exhaustion rather than an improved coordination. This may be caused by inhibitory effects on the pyramidal cells of the brain cortex (Figure 9) as discussed in section 3.3.1 and reflected by depressed magnetically evoked potentials (25). A depressed neuromuscular ex-

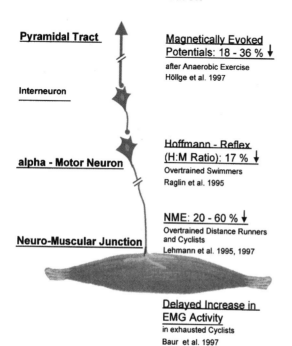

**Overload / Overtraining
Motor Tract / Neuron**

Pyramidal Tract

**Magnetically Evoked
Potentials: 18 - 36 %** ↓
after Anaerobic Exercise
Höllge et al. 1997

Interneuron

alpha - Motor Neuron

**Hoffmann - Reflex
(H:M Ratio): 17 %** ↓
Overtrained Swimmers
Raglin et al. 1995

NME: 20 - 60 % ↓
Overtrained Distance Runners
and Cyclists
Lehmann et al. 1995, 1997

Neuro-Muscular Junction

**Delayed Increase in
EMG Activity**
in exhausted Cyclists
Baur et al. 1997

Figure 11. Schematic presentation of fatigue- or overtraining-related deterioration in motor tract / motor neuron excitability as described by Höllge et al. (25), Raglin et al. (67), Lehmann et al. (47, 48), and by Baur et al. (3).

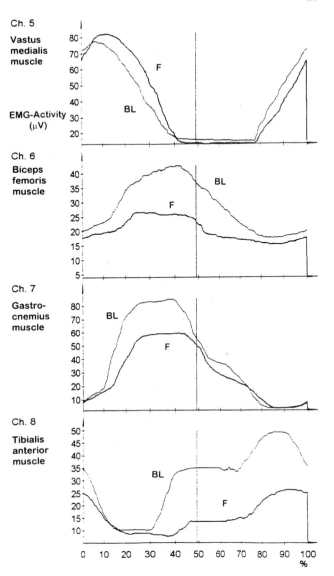

Figure 12. As compared to Vastus medialis muscle, electromyographic measurement revealed (i) a significant right-shift of the increase in electromyographic activity and (ii) a depressed average activity of Biceps femoris, Tibialis anterior, and Gastrognemius muscles at the end of a 1-hour exhausting cycling period (F: final, BL: baseline examination), (3).

citability of the neuromuscular end-plate as discussed in section 2.1 and a depressed excitability of the alpha-motorneuron as discussed in section 3.3.2 likely contribute to this process. But at present, neither alterations in electromyographic activity nor in magnetically evoked potentials have been studied during a state of overtraining-related performance incompetence and recovery from it. Therefore in our opinion, nothing is presently known about the appropriate time span for normalization of these findings subsequent to long-term overtraining.

4. CONCLUSIONS

Findings and mechanisms underlying an overtraining-related performance incompetence can be described in affected athletes as (i) overtraining myopathy, (ii) mild form of sympathetic nervous system insufficiency, (iii) mild form of adrenocortical insufficiency, and (iv) mild form of hypothalamo-pituitary insufficiency in an advanced stage. These findings negatively influence maximum muscular power, maximum energetic competence, motivation and drive thus explaining performance incompetence in affected athletes. As far as known at present, overtraining myopathy is characterized by (i) depressed turnover of contractile proteins particulary of myosin heavy chains of fast twich glycolytic fibers, by (ii) depressed neuromuscular, and by (iii) depressed alpha-motorneuron excitability. The mild form of sympathetic nervous system insufficiency is characterized by (i) depressed ß$_2$-adrenoreceptor density, and by (ii) depressed intrinsic sympathetic system activity. The mild form of adrenocortical insufficiency can be described as a depressed sensitivity of adrenals to ACTH and therefore as depressed maximum cortisol response. The mild form of hypothalamo-pituitary insufficiency is reflected by depressed maximum pituitary adrenocorticotropin and growth hormone responses thus amplifying energetic incompetence in affected athletes. After short-term overtraining of 2–3 weeks, normalization of performance ability can be expected within 1–2 regeneration weeks. However, after long-term overtraining of 3–6 weeks, an extended regeneration period of more than 2 weeks can be expected to be necessary for complete regeneration. But a time span of more than 2 regeneration weeks with a 50–70 % reduction in training load runs the risk of an additional loss in performance ability. This one-dimensional training–regeneration–model is complicated in a completely individual manner by additional competitions and additional non-training stress factors.

REFERENCES

1. Adlercreutz H, Harkonen K, Kuoppasalmi K, Naveri H, Huthamieni H, Tikkanen H, Remes K, Dessipris A, Karvonen J (1986) Effect of training on plasma anabolic and catabolic steroid hormones and their response during physical exercise. Int J Sports Med (Suppl):27–28
2. Barron JL, Noakes TD, Lewy W, Smith C, Millar RP (1985) Hypothalamic dysfunction in overtrained athletes. J Clin Endocrinol Metabol 60:803–806
3. Baur S, Miller R, Liu Y, Freiwald J, Konrad P, Steinacker JM, Lehmann M (1997) Changes in EMG pattern during an hour exhausting cycling. In: Dickhuth HH, Küsswetter W (eds) 35. Deutscher Sportärztekongreß Tübingen. Novartis Pharma Verlag Wehr:265
4. Berg A, Günther D, Keul J (1986) Neuromuskuläre Erregbarkeit und körperliche Aktivität. I. Methodik, Reproduzierbarkeit, Tagesrhythmik. II. Ruhewerte bei Trainierten und Untrainierten. III. Abhängigkeit von beeinflussenden Faktoren. Dtsch Z Sportmed (Suppl) 37:4–22
5. Bieger WP, Zittel R (1982) Effect of physical activity on ß-receptor activity. In: Knuttgen HG, Vogel JA, Poortsmans J (eds) Biochemistry of Exercise. Human Kinetics Publishers Inc, Champaign Il USA:715–722
6. Bruin D (1994) Adaptation and overtraining in horses subjected to increasing training loads. J Appl Physiol 76:1908–1913
7. Brodde ED, Daul A, O'Hara N (1984) ß-adrenoreceptor changes in human lymphocytes, induced by dynamic exercise. Naunyn Schmiedeberg's Arch Pharmacol 325:190–192
8. Budgett R (1990) Overtraining syndrome. Br J Sports Med 24:231–236
9. Butler J, O'Brien M, O'Malley K, Kelle JG (1982) Relationship of ß-adrenoceptor density to fitness in athletes. Nature 298:60–61
10. Costill DL, Flynn MG, Kirwan JP, Houmard JA, Mitchell JB, Thomas R, Sung HP (1988) Effects of repeated days of intensified training on muscle glycogen and swimming performance. Med Sci Sports Exerc 20:249–254
11. Counsilman JE (1955) Fatigue and staleness. Athletic J 15:16–20
12. Foster C, Lehmann M (1996) Overtraining syndrome. In: Guten GN (ed) Running injuries. Saunders Philadelphia:173–188
13. Fry AC (1997) The role of training intensity in resistance exercise overtraining and overreaching. In: Kreider RB, Fry AC, O'Toole ML (eds) Overtraining in sport. Human Kinetics Champaign Il USA:107–127

14. Fry RW, Morton AR, Keast D (1991) Overtraining in athletes. An update. Sports Med 12:32–65
15. Fry RW, Morton AR, Keast D (1992) Periodisation and the prevention of overtraining. Can J Sports Sci 17:241–248
16. Gastmann U, Schiestl G, Schmidt K, Baur S, Steinacker JM, Lehmann M (1995) Einfluß einer BCAA- und Saccharose-Substitution auf Leistung, Aminosäuren- und Hormonspiegel, Blutbild und blutchemische Parameter. In: Kindermann W, Schwarz L (Hrsg) Bewegung und Sport–eine Herausforderung für die Medizin. Ciba Geigy Verlag Basel:165
17. Gastmann U, Schmidt K, Schiestl G, Lormes W, Steinacker JM, Lehmann M (1995) Einfluß einer BCAA- und Saccharose-Substitution auf Leistung, neuromuskuläre Eregbarkeit, EMG und psychometrische Parameter. In: Kindermann W, Schwarz L (Hrsg) Bewegung und Sport–eine Herausforderung. Ciba Geigy Verlag Basel:175
18. Gastmann U ... (1998) Monitoring overload and regeneration in cyclists. This volume
19. Goldberg A (1969) Protein turnover in skeletal muscle. I. Protein catabolism during work induced hypertrophy and growth induced with growth hormone. J Biol Chem 244:3217–3222
20. Goldberg AL, Tischler M, DeMartino G, Griffin G (1980) Hormonal regulation of protein degradation and synthesis in skeletal muscle. Fedn Proc 39:31–36
21. Gruener R, Stern L (1972) Diphenylhydantoin reverses membrane effects in steroid myopathy. Nature 235:54–55
22. Gruener R, Stern L 1972) Corticosteriods: effects muscle membrane excitability. Archs Neurol 26:181–185
23. Hackney AC, Pearman III SN, Nowacki JM (1990) Physiological profiles of overtrained and stale athletes: a review. Appl Sport Psychology 2:21–33
24. Hackney AC, Feith S, Pozos R, Seale J (1995) Effects of high altitude and cold exposure on resting thyroid hormone concentrations. Aviad Space environ Ned 66:325–329
25. Höllge J, Kinkel M, Ziemann U, Tergau F, Geese R, Reimers CD (1997) Central fatigue during exercise. A magnetic stimulation study. In: Dickhuth HH, Küsswetter W (eds) 35. Deutscher Sportärztekongreß Tübingen. Novartis Pharma Verlag Wehr:209
26. Hooper SL, Mackinnon LT, Gordon RD, Bachmann AW (1993) Hormonal responses of elite swimmers to overtraining. Med Sci Sports Exerc 25:741–747
27. Hooper SL, Mackinnon LT, Howard A, Gordon RD, Bachmann AW (1995) Markers for monitoring overtraining and recovery. Med Sci Sports Exerc 27:106–112
28. Houmard JA, Costill DL, Mitchell JB, Park SH, Hickner RC, Roemich JN (1990) Reduced training maintains performance in distance runners. Int J Sports Med 11:46–52
29. Hussar Ü, Seene T, Umnova M (1992) Changes in the mast cell number and the degree of its degranulation in different skeletal muscle fibers and lymphoid organs of rats after administration of glucocorticoids. In: Tissue Biology. Tartu Univeristy Press:7
30. Israel S (1976) Zur Problematik des Übertrainings aus internistischer und leistungsphysiologischer Sicht. Medizin und Sport 16:1–12
31. Jeukendrup AE, Hesselink MKC, Snyder AC, Kuipers H, Keizer HA (1992) Physiological changes in male competitive cyclists after two weeks of intensified training. Int J Sports Med 13:534–541
32. Jost J, Weiss M, Weicker H (1989) Unterschiedliche Regulation des adrenergen Rezeptorsystems in verschiedenen Trainingsphasen von Schwimmern und Langstreckenläufern. In: Böning D, Braumann KM, Busse MW, Maassen N, Schmidt W (Hrsg) Sport, Rettung oder Risiko für die Gesundheit. Deutscher Ärzteverlag Köln:141–145
33. Kelly FJ, Goldspink DF (1982) The differing response of four muscle types to dexamethasone treatment in the rat. Biochem J 208:147–151
34. Kelly FJ, McGrath J, Goldspink D, Gullen M (1986) A morpho-logical (biochemical) study on the actions of corticosteriods on rat skeletal muscle. Muscle Nerve 9:1–10
35. Kreider RB (1997) Central fatigue hypothesis and overtraining. In: Kreider RB, Fry AC, O'Toole ML (eds) Overtraining in sport. Human Kinetics Champaign Il USA:309–331
36. Kreider RB, Fry AC, O'Toole ML (1997) Overtraining in sport. Human Kinetics Champaign Il USA
37. Kuipers H, Keizer HA (1988) Overtraining in elite athletes. Sports Med 6:79–92
38. Lehmann M, Porzig H, Keul J (1983) Determination of ß-receptors on live human polymorphonuclear leukocytes in autologous plasma. J Clin Chem Clin Biochem 21:805–811
39. Lehmann M, Dickhuth HH, Schmid P, Porzig H, Keul J (1984) Plasma catecholamines, ß-adrenergic receptors, and isoproterenol sensitivity in endurance trained and non-endurance trained volunteers. Eur J Appl Physiol 52:362–369
40. Lehmann M, Keul J, Wybitul K, Fischer H (1982) Effect of selective and non-selective adrenoceptor blockade during physical work on metabolism and sympatho-adrenergic system. Drug Res 32:261–266

41. Lehmann M, Foster C, Keul J (1993) Overtraining in endurance athletes. A brief review. Med Sci Sports Exerc 25:854–862
42. Lehmann M, Gastmann U, Steinacker JM, Heinz N, Brouns F (1995) Overtraining in endurance sports. A short overview. Med Sport Boh Slov 4:1–6
43. Lehmann M, Foster C, Netzer N, Lormes W, Steinacker JM, Liu Y, Opitz-Gress A, Gastmann U (1997) Physiological responses to short- and long-term overtraining in endurance athletes. In: Kreider RB, Fry AC, O'Toole ML (eds) Overtraining in sport. Human Kinetics Champaign Il USA:19–46
44. Lehmann M, Foster C, Steinacker JM, Lormes W, Opitz-Gress A, Keul J, Gastmann U (1997) Training and overtraining: Overview and experimental results. J Sport Med Phys Fitness 37:7–17
45. Lehmann M, Gastmann U, Petersen KG, Bachl N, Seidel A, Khalaf AN, Fischer S, Keul J (1992) Training–overtraining: performance and hormone levels, after a defined increase in training volume vs. intensity in experienced middle- and long-distance runners. Br J Sports Med 26:233–242
46. Lehmann M, Baumgartl P, Wiesenack C, Seidel A, Baumann H, Fischer S, Spöri U, Gendrisch G, Kaminski R, Keul J (1992) Training–overtraining: influence of a defined increase in training volume vs. training intensity on performance, catecholamines and some metabolic parameters in experienced middle- and long-distance runners. Eur J Appl Physiol 64:169–177
47. Lehmann M, Jakob E, Gastmann U, Steinacker JM, Keul J (1995) Unaccustomed high mileage vs. high intensity training-related performance and neuromuscular responses in distance runners. Eur J Appl Physiol 70:457–461
48. Lehmann M, Baur S, Netzer N, Gastmann U (1997) Monitoring high-intensity endurance training using neuromuscular excitability to recognize overtraining. Eur J Appl Physiol 76:187–191
49. Lehmann M, Schnee W, Scheu R, Stockhausen W, Bachl N (1992) Decreased nocturnal catecholamine excretion: parameter for an overtraining syndrome in athletes ? Int J Sports Med 13:236–242
50. Lehmann M, Huonker M, Dimeo F, Heinz N, Gastmann U, Treis N, Steinacker JM, Keul J, Kajewski R, Häussinger D (1995) Serum amino acid concentrations in nine athletes before and after the 1993 Colmar ultra triathlon. Int J Sports Med 16:155–159
51. Lehmann M, Knizia K, Gastmann U, Petersen KG, Khalaf AN, Bauer S, Kerp L, Keul J (1993) Influence of 6-week, 6 days per week, training on pituitary function in recreational athletes. Br J Sports Med 27:186–192
52. Lehmann M, Gastmann U, Keizer H, Steinacker JM (1998) Definition, types, symptoms, findings, underlying mechanisms, and frequency of overtraining and overtraining syndrome. This volume
53. Lehmann M, Lormes W, Steinacker JM, Gastmann U (1997) Laboratory markers suitable for training and overtraining state study. In:Österreichische Gesellschaft für Klinische Chemie Wien (ed) Berichte der ÖGKC, ISSN O252–8053, Medical laboratory and sport (Suppl):9–10
54. Liu Y, Opitz-Gress A, Steinacker JM, Zeller C, Mayer S, Lormes W, Gastmann U, Altenburg D, Lehmann M (1997) Adaptation of muscle fibers to 4-week competition preparing training in rowers. In: Dickhuth HH, Küsswetter W (eds) 36. Deutscher Sportärztekongreß Tübingen. Novartis Pharma Verlag Wehr:162
55. Mackinnon LT (1997) Effects of overreaching and overtraining on immune function. In: Kreider RB, Fry AC, O'Toole ML (eds) Overtraining in sport. Human Kinetics Champaign Il USA:219–241
56. Mayr S, Steinacker JM, Opitz-Gress A, Liu Y, Zeller C, Lormes W, Lehmann M (1997) Muscular stress reaction to 4-week competition preparation training in rowers. In: Dickhuth HH, Küsswetter W (eds) 36. Deutscher Sportärztekongreß Tübingen. Novartis Pharma Verlag Wehr:164
57. Morgan WP, Brown DR, Raglin JS, O'Connor PJ, Ellickson KA (1987) Psychological monitoring of overtraining and staleness. Br J Sports Med 21:107–114
58. Naessens G, Lefevre J, Priessens M (1996) Practical and clinical relevance of urinary basal noradrenaline excretion in the follow-up of training processes in semiprofessional soccer players. Clin J Sports Med: in press
59. Netzer N, ... (1998) Sleep and respiration. This volume
60. Nieman DC (1997) Effects of athletic endurance training on infection rates and immunity. In: Kreider RB, Fry AC, O'Toole ML (eds) Overtraining in sport. Human Kinetics Champaign Il USA:193–217
61. O'Toole ML (1997) Overreaching and overtraining in endurance athletes. In: Kreider RB, Fry AC, O'Toole ML (eds) Overtraining in sport. Human Kinetics Champaign Il USA:3–17
62. Owen IR (1964) Staleness. Phys Educ 56:35
63. Parry-Billings M, Bloomstrand E, McAndrew N, Newsholme N, Newsholme EA (1980) A communicational link between skeletal muscle, brain, and cells of the immune system. Int J Sports Med 11 (Suppl 2):122–128
64. Parry-Billings M, Budgett R, Koutedakis Y, Bloomstrand E, Brooks S, Williams C, Calder PC, Pilling S, Baigrie R, Newsholme EA (1992) Plasma amino acid concentration in the overtraining syndrome: possible effects on the immune system. Med Sci Sports Exerc 24:1353–1358

65. Partheniu A, Demeter A (1971) Etude physiologique complexe du quadrceps fémoral, concernant les muscles phasiques et toniques, pendant l'effort phasique chez les élèves. Schweiz Z Sportmed 19:19–30

66. Prellwitz W (1976) Erkrankungen der Nebennierenrinde. In:Prellwitz W (ed) Klinisch-chemische Diagnostik. Thime Stuttgart:311–321

67. Raglin JS, Kocera DM, Stager JM, Harms CA (1996) Mood, neuromuscular function, and performance during training in female swimmers. Med Sci Sports Exerc 28:372–377

68. Rowbotton DG, Keast D, Morton AR (1997) Monitoring and preventing of overreaching and overtraining in endurance athletes. In: Kreider RB, Fry AC, O'Toole ML (eds) Overtraining in sport. Human Kinetics Champaign Il USA:47–66

69. Rowland TW (1986) Exercise fatigue in adolescents: diagnosis of athlete burnout. Phys Sportsmed 14:69–77

70. Salo DC, Donovan CM, Davies KJA (1991) HSP 70 and other possible heat shock or oxidative stress proteins are induced in skeletal muscle, heart, and liver during exercise. Free Rad Biol Med 11:239–246

71. Schneider FJ, Völker K, Liesen H (1993) Zentral- versus peripher-nervale Belastungsreaktion der neuromuskulären Strukturen. In: Tittel K, Arndt KH, Hollmann W (eds) Sportmedizin: gestern–heute–morgen. Barth Leipzig 224–227

72. Seene T, Umnova M, Alev K, Pehme A (1988) Effect of glucocorticoids on contractile apparatus of rat skeletal muscle. J Steroid Biochem 29:313–317

73. Seene T, Alev K (1991) Effect of muscular activity on the turnover rate of actin and myosin heavy and light chains in different types of muscle. Int J Sports Med 12:204–207

74. Seene T (1994) Turnover of skeletal muscle contractile proteins in glucocorticoid myopathy. J Steroid Biochem Molec Biol 50:1–4

75. Snyder AC, Jeukendrup AE, Hesselink MKC, Kuipers H, Foster C (1993) A physiological / psychological indicator of overreaching during intensive training. Int J Sports Med 14:29–31

76. Snyder AC, Kuipers H, Chang BO, Servais R, Fransen E (1995) Overtraining following intensified training with normal muscle glycogen. Med Sci Sports Exerc 27:1063–1070

77. Steinacker JM, Lormes W ... (1998) Findings during overload and regeneration in elite rowers between national and world championships: This volume

78. Stone MH, Keith RE, Kearney JT, Fleck SJ, Wilsond GD, Triplett NT (1991) Overtraining. A review of signs, symptoms and possible causes. J Appl Sport Science Research 5:35–50

79. Tohmeh JF, Cryer PE (1980) Biphasic adrenergic modulation of ß-adrenergic receptors in man. J Clin Invest 65:836–840

80. Urhausen A, Gabriel HHW, Kindermann W (1998) Impaired pituitary hormonal response to exhaustive exercise in overtrained athletes. Med Sci Sports Exerc 30:407–414

81. Verma SK, Mahindroo SR, Kansal DK (1978) Effect of four weeks of hard physical training on certain physiological and morphological parameters of basketball-players. J Sports Med 18:379–384

82. Wittert GA, Livesey JH, Espiner EA, Donald RA (1996) Adaptation of the hypothalamopituitary-adrenal axis to chronic exercise stress in humans. Med Sci Sports Exerc 28:1015–1019

3

PERSPECTIVES ON CORRECT APPROACHES TO TRAINING

Carl Foster, [1*]Jack T. Daniels, [2] and Stephen Seiler[3]

[1]Department of Exercise and Sport Science
132 Mitchell Hall, University of Wisconsin-LaCrosse
LaCrosse, Wisconsin 54601
[2]SUNY-Cortland
Cortland, New York
[3]Agder College
Kristiansand, Norway

1. INTRODUCTION

The life of endurance athletes is about effort, effort during competition and especially effort during training. In preparation for an Olympic competition in the marathon, a race of about 2 hours, an elite runner may train 20 hours per week for nine months or more. In other sports such as cycling, swimming or Nordic skiing, the number of training hours may exceed this. In individuals willing to spend the time and effort to do the requisite training, remarkable endurance performances are possible. This ability to improve with training presents a seductive lure to do more and more training in quest of achieving a desired competitive result. However, it is clear that without the basic ability to perform at a good level with minimal training, no amount of training will "make a thoroughbred out of a plow horse".

Endurance sports attract athletes and coaches who are effort oriented. They are more than willing to push their efforts during competition to the point of physical collapse. They are also more than willing to "train to a standstill", in the belief that such extreme levels of training will result in improved performances. Many of the great champions in a variety of endurance sports have been individuals with extraordinary ability to tolerate levels of training that are significantly beyond that of their contemporaries; levels of training that prove devastating when other athletes attempt to replicate the champions program.

* Phone 608 785 8687, FAX 608 785 8172, e-mail foster@mail.uwlax.edu

Overload, Performance Incompetence, and Regeneration in Sport, edited by Lehmann *et al.*
Kluwer Academic / Plenum Publishers, New York, 1999.

Additionally, endurance athletes (and their coaches) are almost inevitably predisposed to respond to failure, whether a bad training session or a bad competition, with increased effort. Thus, the nature of competition, of the training done in preparation for competition, and the type of individuals attracted to endurance sports predisposes toward overtraining syndrome as much as towards world records.

There is a rich literature concerning training for endurance sports. It seems that endurance athletes like to read about training almost as much as they like to train. However, most of this literature is phenemenological in nature. It reports what worked for a particular athlete or group of athletes. Some of this literature is quite good, synthesizing the results of years of informal "experimentation" by coaches (1, 12, 28, 33) or insightful syntheses by informed observers (36, 40).

The unfortunate reality is that it is difficult to do controlled experimental trials with elite athletes. Most of the systematically collected data are based on either recreational or sub-elite athletes, often observed for periods of time that are unrepresentatively short compared to the buildup periods of elite athletes. Nevertheless, there remains a nucleus of systematically collected data that least begin to suggest how to use training, not only to prevent injuries and overtraining syndrome, but to achieve the best performance. Accordingly, this chapter is intended to evaluate the available data on the training of athletes in endurance sports with the goal of presenting a testable hypotheses regarding "correct training".

2. TRAINING CHARACTERISTICS

Every training program represents a complex interplay of a variety of factors. However, the simplest way to evaluate how training might relate both to improved performance and to the risk side effects is to break training into various constituents such as volume, intensity, and specificity of training. However, even as we do this, we must remember that responses to training (both positive and negative) are highly complex and individual.

2.1. Training Volume

Endurance athletes have a very strong bias toward training volume. At the most simple level, all one has to do is ask any endurance athlete how hard they are training and you will receive an answer based on kilometers per week, laps per day or some similiar indication of training volume. It is fair to say that the volume of training is widely considered the central issue, with intensity training important primarily as a sharpening tool.

In very long distance competitions, such as marathon running, there is the concept of the "collapse point" (first suggested by the statistician/runner Ken Young 50). This is the distance the athlete can run without needing to walk for recovery, or the minimal distance one needs to be training to complete the event in "good order". Obviously, a greater volume of training may be required to compete optimally. The collapse point hypothesis proposes that the weekly training volume needs to be at least 2.5 times the racing distance. Thus, a marathon runner would need to be averaging at least 105 km per week in order to complete the distance. While there are no directly comparable volume guidelines in other sports, it would be perfectly ordinary for an elite cyclist to cover 300 km per week in training in preparation for a race of 120 km. Even for longer distance events, such as the 90 km Comrades marathon, the competitior would need to be averaging 225 km per week to get through the event in "good order" (a training volume not unrealistic for elite competitors). At the same time, there are many empirical suggestions that these kinds of training volumes are not necessary (50), particularly in the ultra-distance events. At the other end of the continuum, a competitor just interested in finishing a 1.5 km in good order would need to be training only 3.75 km per week. This value is clearly unrealistic. Thus, one may propose that the minimal training volume required to complete an event in "good order" is a curve, with an empirically justified fixed point of 2.5* race duration for events of about 3 hours in duration. The shape of the curve, needs to be experimentally verified, but may approximate that presented in Figure 1. The alternative hypothesis would be that the "collapse point" concept is only meaningful in competitions exceeding 90–120 minutes where muscle glycogen depletion is a primary limiting factor.

There are several observational cross sectional studies which address the issue of training volume in relation to running performance in the 10 km to marathon. On the basis of questionnaires designed to document running performances in shorter distances (as an index of general running ability) and training over the last eight weeks before competition, several authors (15–17, 24, 25) have noted the importance of training volume (Figure 2). Particularly in the marathon, training volume is significantly related to performance. Analysis of the slowdown during the latter part of marathon races provides additional support for the collapse point hypothesis (15). In less experienced competitors, the number of

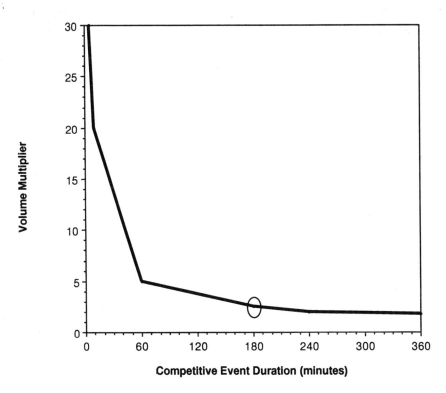

Figure 1. Schematic relationship between the duration of the event being prepared for and the relative volume of training necessary to complete the event in good order. For events similiar to recreational class marathon running (~3 hours), there is experimental support for a multiplier of about 2.5 (circle). For events of shorter or longer durations, there is no experimental evidence for a particular multiplier. The values presented here are empirically derived.

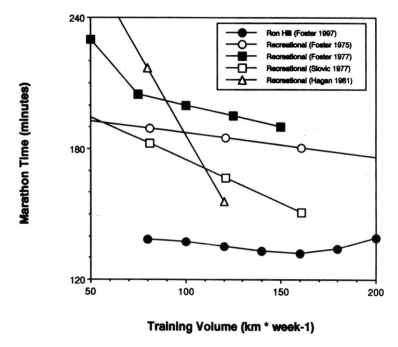

Figure 2. Schematic relationship between training volume during the two months preceding competition and marathon running performance. Although some of the cross-sectional studies (open symbols) suggest a large effect, the two longitudinal studies (closed symbols) suggest that each additional 10 km per week in training is only worth 1–2 minutes in performance. There is, however, evidence that at very low training volumes (<80 km per week) there is a larger effect from additional training volume. Similarly, at very high training volumes (>160 km per week) there may be a negative effect from additional training volume.

long training runs (e.g. more specific training) seemed to be as critical as training volume. However, even in distances as short as 10 km, training volume has been shown to be of importantce (16). Sjodin (42, 44) has noted that marathon performance, the running velocity at a blood lactate concentration of 4mmol*l-1, and the activity of selected skeletal muscle enzymes are all associated with total distance run per week. However the relationship flattens significantly above 140 km per week. On the other hand, Scrimgeour et al (39) noted that there was no relationship between training volume and the %VO2max which could be sustained during standard competitions.

There are few longitudinal data regarding the effect of training volume per se on endurance performance. We have followed the training of a group of recreational marathon runners through the course of several marathons (17). There was a generalized trend for improved performance with greater training volume (Figure 1). However, it may be that once a critical training volume is achieved that subsequent increases are counterproductive. We (21) have performed an analysis of the training of the British marathoner Ron Hill based on data provided by Noakes (36). Hill was legendary for his committment to training, and unfortunately also legendary for his tendency to fail in big competitions. The data suggest that once Hill exceeded about 160 km per week, his performance deteriorated (Figure 2). The results of these two studies suggests that once a training threshold of about 100 km per week is achieved, that the effect of additional increases in training volume are small (on the order of 1–2 minutes faster for each additonal 10 km per week in training).

These observations fit well with the development of overtraining syndrome in athletes participating in the increased training volume study of Lehmann et al. (30, 31), and suggest that the margin between the training necessary to get beyond the collapse point and that which is tolerable without side effects becomes quite small at the level of training undertaken by serious athletes.

2.2. Training Intensity

Although endurance athletes tend to focus on the importance of training volume, the more important factor in the training of athletes is probably training intensity. Human competitive athletes are not migratory animals. The prize is given not for continuing longer, but for going faster than one's competitors. Clearly, practice must be organized to allow the athlete to adapt to the specific demands of the intensity of competition, particularly the muscle fiber recruitment pattern specific to the competitive event. Higher intensity training is designed to accomplish this goal. At the same time that the value of higher intensity training is recognized; most authorities suggest that no more than 5–20% of the training load be accomplished at intensities greater than the anaerobic threshold (1, 12, 27, 36, 40). Despite the failure to develop overtraining syndrome in the one well done study of additional high intensity training (30, 31), there is a wealth of practical experience which suggests that a large volume of high intensity training is poorly tolerated.

There are several cross sectional studies suggesting the importance of higher intensity training (≥ "anaerobic threshold") on performance. In a comparison of elite black and white South African runners, Coetzer et al (7) noted that the black athletes were able to run at a higher %VO2max during races > 5 km, and that a higher percentage of their training was at high intensity (36% vs 14%) versus white athletes, with both groups having similar training volumes. Importantly, many of the black athletes have a many year history of a high volume of low intensity training as part of the requirements of living in their particular society. Using an index of training intensity, we (18) have noted significant correlations with performance and the presence of high intensity training, particularly in the shorter endurance racing events (≥5km).

Contemporary coaching opinion also suggests that, at least at certain times of the preparatory period, high intensity training is absolutely essential toward the development of athletes (12, 28, 33, 40). Both Daniels (12) and Seiler (40) suggest that the program is built around moderate to higher intensity long interval training, with everything else forming a supporting role.

There are several longitudinal studies which support the importance of training intensity. In an observational case study of the American 1500m runner, Steve Scott, Conley et al. (8) noted that performance, VO2max, running economy and body compositon all improved during the period of the year when the quantity of interval training increased. Gorostiaga et al. (23) noted improved performances and VO2max in a group performing interval training compared to a continuous training group (training at the same average power output). Hickson et al. (25) noted that maintence of training intensity was critical to maintaining performance during experimental reductions in training load. Neither training duration or frequency seemed particularly critical. Lehmann et al. (30, 31) noted improved performance during an experimental study of increased training intensity, and apparent resistance to overtraining syndrome, possibly because of the hard day-easy day adopted during this study.

In one of the more widely cited experimental training studies, Sjodin et al. (43) studied runners who added a single 20 minute training session at the intensity associated with

OBLA to their ordinary training (which was predominately easier than the intensity associated with OBLA). There was a subsequent increase in the velocity associated with
OBLA following training. Interestingly, the closer to a steady state of blood lactate the
athletes trained, the greater the subsequent increase in OBLA after training. This finding
has been taken to suggest that the training intensity associated with the maximal lactate
steady state may be "optimal" for improving endurance performance. This finding is in
substantial agreement with Hollman who noted significant increases in OBLA following
training at the intensity associated with OBLA, but no change following training at 95%
VO2max (27).

The suggestion that the intensity associated with the maximal lactate steady is a
uniquely ideal training intensity is challenged, however, by the results of Keith et al. (29)
who noted equivalent improvement in groups training using steady state exercise at the individual anaerobic threshold versus interval training at the same average power output. Although failing to achieve statistical significance, the increase in endurance time at the
power output associated with the pre-training individual anaerobic threshold actually improved more in the interval training group than in the continuous training group (52% vs
43%).

In a recent, well designed and most practical study, Lindsay et al. (32) evaluated the
effect of adding interval training (5 minute repetitions at 80% of the maximal incremental
power output) to the training of already well trained cyclists. Particular care was taken to
maintain the same total volume of training, with the additional interval training replacing
15% of the pre experimental training volume. Preceeding studies had demonstrated the
stability of performance in these athletes. Over the course of four weeks, cycle time trial
performance (40 km) improved by two minutes (3.5%). Considering the stability of performance before the addition of interval training, and the reality that the competitively
meaningful range of performances is less than 2%, the results of this study support the importance of higher intensity training. Another recent study from the same group, using a
unique modeling approach to evaluate improved cycle time trial performance, demonstrated improved performances following some versions of added interval training but not
others (41). The results were not uniquely supportive of more specific training, but sug-
gestive of highly individual response patterns to interval training.

Training intensity may, however, be only of modest importance within the overall
scheme of increases in trainining load. Following a preliminary baseline training period,
Daniels et al. (11), increased the training load in both novice and experienced runners.
Some of the subjects increased their training loads with a greater volume of steady state
running, while some performed increased amounts of interval running. There did not seem
to be a strong relationship between how the training load was increased and the magnitude
of improvement in performance over both 800m or 3200m.

In the absence of definitive data, it seems prudent to accept the implications of the
published data, and the empirical wisdom of experienced coaches, and suggest that training intensity is probably the critical element of training programs. Rather than suggesting
that one training intensity is ideal for every athlete, it seems more reasonable to suggest
(on the basis of the muscle fiber recruitment pattern) that the majority of higher intensity
training should be approximate the intensity contemplated for impending competitive
events. At the same time, the volume of training which may be accomplished at high intensity is probably limited. In the absence of definitive data, probably less than 20% of the
total training volume should be accomplished at high intensity. A schematic of exercise intensity in relation to the generally recommended percentage of training time in various
training zones is presented in Figure 3.

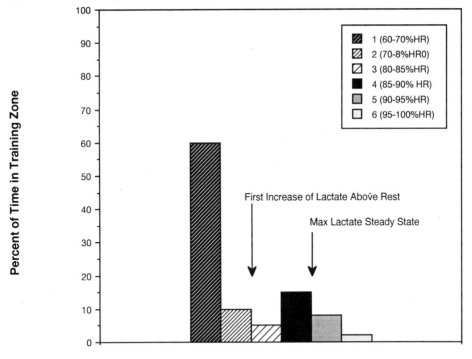

Training Zone

Figure 3. Schematic depiction of the relative percentage of training time in different training intensity zones. Note that most training (60–70%) is widely thought to be at training intensities with no net increase in the blood lactate concentration, and that less than 10% of training above the maximal lactate steady state is thought to be desirable.

2.3. Training Load

As much as it is experimentally tidy to discriminate between training volume and training intensity, the real world reality is that athletes mix both elements into their training. Banister et al. (14, 34), in the process of developing a systems model of the response to training, pioneered the concept to the training impulse (TRIMP) or the product of training duration and intensity (calculated as average heart rate multiplied by a non linear metabolic adjustment multiplied by the duration of effort). They demonstrated that both changes in fitness (a positive response to training) and fatigue (a negative response to training) could be calculated and could account for variations in performance. This concept has been extended by Busso et al. (5, 6) and Mujika et al. (35) in France, using a rating system based on percentages of maximal possible performance to estimate exercise intensity, with satisfying results at least in relation to developing a model of the response to training.

We (20) have used a modification of the rating of perceived exertion scale as a marker of global exercise intensity. Multiplying the RPE for the entire exercise session by the duration of each training session gives an index comparable to Bannister's of the training impulse for that session. Using this technique, when the weekly load is averaged over eight weeks, there is a good relationship between training load and performance, with a ten fold increase in training load being associated with about a 10% improvement in a cy-

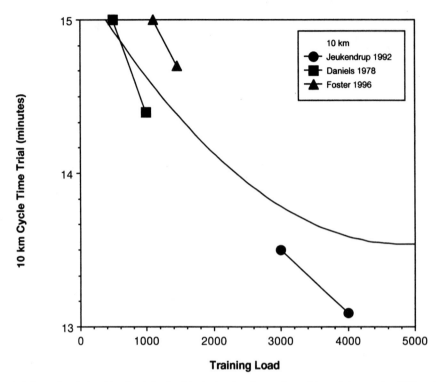

Figure 4. Schematic relationship of training load (calculated as the product of the training session RPE and training duration (ref 20)) and expected performance in a 10 km cycling event (curve). Note that as the training load is increased 10 fold from basic fitness levels of training (500 units per week), the time required improves by about 10%. Data from comparable experimental training studies are plotted for comparison.

cle 10 km time trial (Figure 4). This training induced improvement in performance fits other experimental observations for events of similiar durations. For marathon length events, the magnitude of improved performance in relation to increases in training load was larger.

Our index of training intensity correlates well with heart rate estimates of global exercise intensity and with blood lactate derived training zones (22). It also has the advantage of not being technologically dependant and of being able to account for very high intensity training (such as resistance training) where heart rate might not provide a valid index of intensity.

Beyond the preliminary data available using indices of training load, the larger importance here is that there are a variety of methods of combining training volume and intensity into a single number. This opens the possibility of making some more representative observational studies, or even realistically controlled studies to determine the quantitative relationship between training and performance.

2.4. Specificity

It is widely agreed that training should be as specific as possible. Thus, despite the spate of recent studies demonstrating that cross training (non specific training) can maintain general fitness during periods of vacation or injury (3, 26, 37, 48, 49), the more specific the training the larger the training effect. In a recent study we compared the value of adding additional running training versus adding swimming training to baseline levels of running in a group of recreational runners (19). With the total training load equated, the subsequent improvement in performance was about twice as much in the specifically trained group as in the cross training group, with no change in a control group maintaining baseline levels of running. This finding is consistent with observations demonstrating that training effects are very specific to the musculature which is trained (38) and that training effects transfer much better to muscularly similar types of activity (3, 26, 37, 48, 49). A combined analysis of both muscularly similar and muscularly disimilar cross training suggests that muscularly similar training has about 75% and muscularly dissimilar training has about 50% of the value of specific training on a minute by minute basis.

In the interval training study of Lindsay et al (32), the experimentally added interval training was performed at about the same intensity (80% PPO) as the criterion performance (40 km time trial). Thus, it may be that it was the more specific training rather than the higher intensity per se that contributed to the improved performance. However, in companion experiments comparing the influence of moderate intensity endurance training and high intensity intermittent training in previously moderately trained subjects, Tabata et al. (47) observed that both training programs resulted in an increase in aerobic power, but that only the high intensity intermittent training was associated with increases in the maximal accumulated oxygen deficit. This suggests that the effect of high intensity training may operate via a mechanism other than the aerobic power. Similarly, Stepto et al. (41) observed improved cycle time trial performance in already well trained cyclists after slightly more intense and very much more intense exercise bouts, but not at very specific exercise bouts. They suggested, that there may be an individual responsiveness to various types of interval training that is larger than a specific response.

2.5. Interval Training

Despite some of the problems with specificity noted above, it is probably true that training should be as specific as possible, and it is also agreed that the metabolic intensity of exercise should not be overly high. Accordingly, interval training was developed a generation or more ago to allow athletes to do a large volume of comparatively high intensity, muscularly specific training with only modest global disturbances in homeostasis. A variety of studies have demonstrated that if the "hard" periods of interval training are kept comparatively short (20–60s), that fairly high intensity exercise may be performed well within the range of the maximal lactate steady state. The longer the "hard" period of work, the greater disturbance to homeostasis. The length of the recovery period between intervals is comparatively less important. As a generic approach, it is interesting to note that independently developed syntheses of training in several different sports, suggest the value of repeating longer intervals (2–3 minutes) at the velocity associated with the maximal lactate steady state (12, 40).

2.6. Periodization of Training

There is a very large coaching literature on the periodization of training, or how to mix various types of training to get an optimal competitive result at the desired time. In his classic work for distance runners, Lydiard (33) presents the problem well...paraphrased..."I found that I could easily get runners to perform at very high levels, but it took many years for me to find the training pattern that allowed them to reach peak performance on the day they wanted peak performance". Unfortunately, for all the collective wisdom of experienced coaches (1, 12, 28, 33, 36), there are virtually no controlled data addressing the value of training periodization. Even the longitudinal intervention studies with various types of athletes have used intervention periods of only 6–10 weeks, and have not focused on the importance of training done prior to an intervention period. Our understanding of training periodization is indeed, based much more on empirical observation, and on creative conjecture than on data.

As part of their attempt to develop a systems model of the response to training, Bannister et al. (14, 34) have created the concept of TRIMPS and the interplay between fitness and fatigue. In their concept, each training session provokes both an increase in fitness and an increase in fatigue. For a training session sufficient to momentarily suppress performance ability, the subsequent recovery period will allow for improved performance because residual fatigue resulting from training fades more rapidly (time constant of about 10 days) than the residual fitness resulting from training (time constant of about 40 days). Ultimately studies of this nature should provide some objective evidence regarding various periodization schemes.

In an attempt to understand factors which might relate to some of the banal illnesses that are associated with overreaching and overtraining syndrome, we (21, 22) have developed the monotony hypothesis. The less variable (=more monotonous) the training load, the more likely athletes seem to be to develop banal illnesses. This occurs regardless of whether a particular training load is accumulated by extensive low intensity training or by brief intensive training. These data fit well with animal models of severe training (2). Whether more training variability (=less monotonous) on a day to day basis might also lead to better adaptive responses, remains to be determined. However, given the propensity of endurance ahtletes toward effort, it may be that the best "recovery" training is total rest.

3. TRAINING ADJUNCTS—ALTITUDE TRAINING

Training occurs within the context of other environmental factors which may either facilitate or hinder subsequent endurance performance. Clearly, when preparing for competition in extreme environmental circumstances (heat, cold, high altitude), most of the training should be in the conditions likely to be encountered during competition. Collective evidence suggests that it may take 4–8 weeks for the first phase of adaptation to heat/humidity or high altitude to take place, with natives always appearing to have some competitive advantage. There is less data on acclimaztion to severe cold, but it surely occurs.

Based on studies demonstrating the value of induced polycythemia on endurance performance (13), there has been much interest in strategies that might be useful to effectively increase the oxygen carrying capacity of the blood. Both blood reinfusion and rEPO administration are unethical, as well as being associated with significant risk of concominant morbidity/mortality. Both early studies of altitude training (10) and contemporary studies of the live high-train low hypothesis (46) suggest that sustained altitude exposure (>2300m) may induce polycythemia, may improve aerobic physiologic capabilities and may result in subsequent improvement in sea level performances. Apparently, at least some high intensity training needs to be done at sea level in order to prevent loss of muscle power during altitude exposure. The implication of these studies is very clear. Altitude exposure is an important, independent, mechanism of enhancing the response to training.

4. CONCLUSION

At the present time, we simply don't know enough about the correct way to train to present a definitive set of guidelines that will work for all athletes in all endurance disciplines. The best that may be offered is to point out the similiarity in recommendations in independently derived syntheses of how to train.

Arthur Lydiard, the New Zealand coach of gold medalists Murray Halberg and Peter Snell, and the influencer of a generation of athletics coaches makes several basic recommendations (33):

1. You can't train hard and race hard at the same time.
2. If you can run more than 100 miles per week (161 km) during the build-up period, don't run more, run faster.
3. 100 miles per week is relatively easy, 20 miles one day, 10 miles the next day
4. Once the base training is done, success is related to a controlled buildup of training intensity.

In his classic work Lore of Running (36), Noakes presents an insightful synthesis of the training programs of several elite long distance runners or coaches of elite runners. He presents this synthesis in the form of 15 Laws of Training. These laws may be summarized as:

1. Train consistently
2. Start gradually and train gently
3. Train first for distance, later for speed
4. Don't be overly compulsive about a daily schedule
5. Alternate hard and easy training
6. Try to achieve as much as possible on minimal training
7. Don't race when training, run time trials and races infrequently
8. Train specifically
9. Use Base training and Sharpening
10. Don't overreach
11. Train with a coach
12. Train the mind
13. Rest before major competitions
14. Keep a detailed training log
15. Understand the holism of training

Jack Daniels, a highly successful US athletics coach makes several basic suggestions that seem to apply to both elite and developmental runners (12).

1. Don't leave your race on the training track.
2. Alternate hard and easy days, in fact only 2–3 hard days per week.
3. Improvement is built around the judicious use of intensity (repetition training and threshold training).
4. If recovery training doesn't leave you recovered, try resting entirely.

Stephen Seiler, an American ex-patriate living in Norway, who has developed an interesting synthesis of training based on experiences with cyclists, rowers, and cross country skiers has several basic principles (40).

1. Build the program around two high intensity interval sessions per week.

2. Most of the non-interval training should be at fairly low intensities (below the first increase in the blood lactate concentration).
3. Avoid the middle intensities (slightly increased blood lactate concentrations), they do little more than tire you out.
4. If you aren't training easily enough on the easy days, you won't be able to train hard enough on the hard days.
5. The total volume of low intensity "background" training serves as a platform for progressively higher intensity "specific" training. Thus an athlete who can tolerate 15 hours per week of background training will be able to tolerate higher intensity specific training than an athlete who can only tolerate 10 hours per week of backrougnd training. The limits of tolerability of background train are: a) specific to the sport with cycling/swimming at the upper end and running at the lower end, b) developed over months and years of training, and c) highly variable amongst individuals. This is depicted in Figure 5.

What is remarkable about these independent approaches to training is the similiarity and many points of agreement. They form the basis for developing several testable hypotheses regarding training.

1. The minimal volume of training necessary to compete effectively throughout the duration of an event may be described by a curvilinear function, relating the ex-

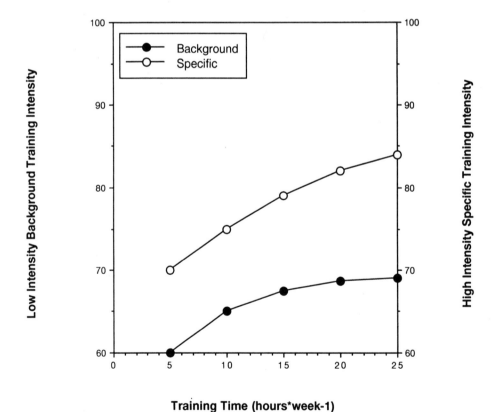

Training Time (hours*week-1)

Figure 5. Schematic relationship of the relative intensity of background training and higher intensity specific training to either the total training load or to progresssion over successive years of training.

pected DURATION of the competitive event to a multiplier. For competitions of about 3 hours duration, the multiplier is about 2.5 (or 7.5 hours weekly). For shorter duration events, the multiplier is necessarily larger, for longer duration events, the multiplier is smaller. The numerial value of the duration multiplier must be determined experimentally.

2. Training intenstiy is probably the single most important factor in correct training. Just as there is a hypothesized relationship between the total volume of training and the duration of the proposed competitive event, there is a proposed curvilinear relationship between the volume of higher intensity (e.g. specific race pace) training and performance. The numerical relationship between the volume of highly specific training and race duration must be determined experimentally.

3. More variation in the day to day training load is likely to result in more effective adaptation to training. The exists and experimentally determinable "best mixture" between the training load from hard to easy days.

4. The adaptive response to the addition of specific training is quantitatively dependant upon the total volume of training performed during the build up period.

REFERENCES

1. Bowerman WJ, Freeman WH (1991) High Performance Training for Track and Field. Champaign, IL, Leisure Press.
2. Bruin G, Kuipers H, Keizer HA, Vander Vusse GJ (1994) Adaptation and overtraining in horses subjected to increasing training loads. J Appl Physiol 76: 1908–1913.
3. Bushman BA, Flynn MG, Andres FF, Lambert CP, Taylor MS, Braun WA (1997) Effect of 4 wk of deep water run training on running performance. Med Sci Sports Exerc 29: 694–699.
4. Busso T, Carasso C, Lacour J-R (1991) Adequacy of a systems structure in the modeling of training effects on performance. J Appl Physiol 71: 2044–2049.
5. Busso T, Candau R, Lacour J-R (1994) Fatigue and fitness modelled from the effects of training on performance. Eur J Appl Physiol 69: 50–54.
6. Busso T, Dennis C, Bonnefoy R, Geyssant A, Lacour J-R (1997) Modeling of adaptations to physical training by using a recursive least squares algorithm. J Appl Physiol 82: 1685–1693.
7. Coetzer P, Noakes TD, Sanders B, Lambert MI, Bosch AN, Wiggins T, Dennis SC (1993) Superior fatigue resistance of elite black South African distance runners. J Appl Physiol 75: 1822–1827.
8. Conley DL, Krahenbuhl GS, Burkett LN (1981) Training for aerobic capacity and running economy. Physician Sportsmed 9: 107–114.
9. Costill DL, King DS, Thomas R, Hargreaves ÊM (1985) Effects of reduced training on muscular power in swimmers. Physician Sportsmedicine 13: 94–101.
10. Daniels J, Oldridge N (1970) The effects of alternate exposure to altitude and sea level on world class middle distance runners. Med Sci Sports 2: 107–112.
11. Daniels JT, Yarbrough RA, Foster C (1978) Changes in VO2max and running performance with training. Eur J Appl Physiol 39: 249–254.
12. Daniels J (1998) Daniels' Running Formula Champaign, IL, Human Kinetics Press
13. Ekblom B (1976) Central circulation during exercise after venesection and reinfusion of red blood cells. J Appl Physiol 40: 379–384.
14. Fitz-Clarke JR, Morton RH, Banister EW (1991) Optimizing athletic performance by influence curves. J Appl Physiol 71: 1151–1158.
15. Foster C, Daniels J (1975) Running by the numbers. Runners World, July.
16. Foster C, Daniels J (1975) Winning cross country. Runners World, Sept.
17. Foster C, Daniels JT, Yarbrough RA (1977) Physiological and training correlates of marathon running performance. Aust J Sports Med 9: 58–61.
18. Foster C (1983) VO2max and training indices as determinants of competitive running performance. J Sports Sci 1: 13–22.

19. Foster C, Hector L, Welsh R, Schrager M, Green MA, Snyder AC (1995) Effects of specific versus cross training on running performance. Eur J Appl Physiol 70: 367–372.

20. Foster C, Daines E, Hector L, Snyder AC, Welsh R (1996) Athletic performance in relation to training load. Wisc Med J 95: 370–374.

21. Foster C, Lehmann M (1997) Overtraining syndrome. In: Guten GN (ed) Running Injuries, Philadelphia, WB Saunders Co: 173–188.

22. Foster C (1998) Monitoring training in ahletes with reference to indices of overtraining syndrome. Med Sci Sports Exerc 30: 1164–1168.

23. Gorostiaga EM, Walter CB, Foster C, Hickson RC (1991) Uniqueness of interval and continuous training at the same maintained exercise intensity. Eur J Appl Physiol 63: 101–107.

24. Hagan RD, Smith MG, Gettman LR (1981) Marathon performance in relation to maximal aerobic power and training indices. Med Sci Sports Exerc 13: 185–189.

25. Hickson RC, Foster C, Pollock ML, Galassi TM, Rich S (1985) Reduced training intensities and loss of aerobic power, endurance and cardiac growth. J Appl Physiol 58: 492–499.

26. Hoffmann JJ, Loy SF, Shapiro BI, Holland GJ, Vincent WJ, Shaw S, Thompson DL (1993) Specificity effects of run versus cycle training on ventilatory threshold. Eur J Appl Physiol 67: 43–47.

27. Hollmann W, Rost R, Liesen H, Dufaux B, Heck H, Mader A (1981) Assessment of different forms of physical activity with respect to preventive and rehabilitatiove cardiology. Int J Sports Med 2: 67–80.

28. Holum D (1984) The Complete Handbook of Speed Skating, Hillside, NJ, Enslow Publishers, Inc..

29. Keith SP, Jacobs I, McLellan TM (1992) Adaptations to training at the individual anaerobic threshold. Eur J Appl Physiol 65: 316–323.

30. Lehmann M, Dickhuth HH, Gendrisch G, Lazar W, Thum M, Kuminski R, Aramendi JF, Peterke E, Wieland W, Keul J (1991) Training-overtraining: A prospective, experimental study with experienced middle and long distance runners. Int J Sports Med 12: 444–452.

31. Lehmann M, Baumgartl P, Wiesenack C, Seidel A, Baumann H, Fischer S, Spori U, Gendrisch G, Kaminski R, Keul J (1992) Training-overtraining: influence of a defined increase in training volume vs training intensity on performance, catecholamines and some metabolic parameters in experienced middle and long distance runners. Eur J Appl Physiol 64: 169–177.

32. Lindsay FH, Hawley JA, Myburgh KH, Schomer HH, Noakes TD, Dennis SC (1996) Improved athletic performance in highly trained cyclists after interval training. Med Sci Sports Exerc 28: 1427–1434.

33. Lydiard A, Gilmore G (1962) Run to the Top, Wellington, NZ, A.H. and A. Reed Publishers.

34. Morton RH, Fitz-Clarke JR, Banister EW (1990) Modeling human performance in running. J Appl Physiol 69: 1171–1177.

35. Mujika I, Busso T, Lacoste L, Barale F, Geyssant A, Chatard J-C (1996) Modeled responses to training and taper in competitive swimmers. Med Sci Sports Exerc 28: 251–258.

36. Noakes TD (1991) Lore of Running, 3rd edition, Champiagn, IL, Human Kinetics Publishers.

37. Pizza FX, Flynn MG, Starling RD, Brolinson PG, Sigg J, Kubitz ER, Davenport RL (1995) Run training vs cross training: Influence of increased training on running economy, foot impact shock and run performance. Int J Sports Med 16: 180–184.

38. Saltin B: Nazar K, Costill DL, Stein E, Jansson E, Essen B, Gollnick PD (1976) The nature of the training response; peripheral and central adaptations to one-legged exercise. Acta Physiol Scand 96: 289–305.

39. Scrimgeour AG, Noakes TD, Adams B, Myburgh K (1986) The influence of weekly training distance on fractional utilization of maximum aerobic capacity in marathon and ultramarathon runners. Eur J Appl Physiol 55: 202–209.

40. Seiler S (1997) Endurance training theory-Norwegian style. Internet http:www.krs.hia.no/~stephens/-xctheory.htm

41. Septo NK, Hawley JA, Dennis SC, Hopkins WG (In Press) Effects of different interval training programs on cycling time trial performance in well trained men. Med Sci Sports Exerc.

42. Sjodin B, Thorstensson A, Frith K, Karlsson J (1976) Effect of physical training on LDH activity and LDH isozyme pattern in human skeletal muscle. Acta Physiol Scand 97: 150–157.

43. Sjodin B, Jacobs I, Svedenhag J (1982) Changes in onset of blood lactate accumulation (OBLA) and muscle enzymes after training at OBLA. Eur J Appl Physiol 49: 45–57.

44. Sjodin B, Svedenhag J (1985) Applied physiology of marathon running. Sports Med 2: 83–99.

45. Slovic P (1977) Empirical study of training and performance in the marathon. Res Quart 48: 769–777.

46. Stray-Gundersen J, Chapman JR, Levine BD (1998) HILO training improves performance in elite runners. Med Sci Sports Exerc 30: S35.

47. Tabata I, Nishimura K, Kouzaki M, Hiral Y, Ogita F, Miyachi M, Yamamoto K (1996) Effects of moderate intensity endurance and high intensity intermittent training on anaerobic capacity and VO2max. Med Sci Sports Exerc 28: 1327–1330.

48. Tanaka H (1994) Effects of cross training: Transfer of training effects on VO2max between cycling, running and swimming. Sports Med 18:330–339.
49. Wilber RL, Moffatt RJ, Scott BE, Lee DT, Cucuzzo NA (1996) Influence of water run training on the maintence of aerobic performance. Med Sci Sports Exerc 28: 1056–1062.
50. Young K (1973) The theory of collapse. Runners World, Sept.

4

MONITORING OF TRAINING, WARM UP, AND PERFORMANCE IN ATHLETES

Carl Foster,[1*] Ann Snyder,[2] and Ralph Welsh[2]

[1]Department of Exercise and Sport Science
132 Mitchell Hall
University of Wisconsin-LaCrosse
LaCrosse, Wisconsin 54601
[2]University of Wisconsin
Milwaukee, Wisconsin

1. INTRODUCTION

Overtraining syndrome is a serious disorder, equivalent in severity to many orthopaedic injuries, and often sufficient to end a competitive season. Although various therapeutic approaches have been tried, overtraining syndrome is generally refractory to treatments other than an extended rest from heavy training and competition. Accordingly, prevention of overtraining syndrome is of critical importance. Although widely studied, the ultimate causes and pathophysiologic nature of overtraining syndrome are not fully understood (4, 7–9). There is a general understanding of the factors likely to cause overtraining syndrome relative to the structure of the training program, with large increases in training load, training monotony, travel, frequent competition and social factors all thought to increase the liklihood of developing overtraining syndrome. Despite extensive study, the diagnosis of overtraining syndrome still remains a diagnosis by exclusion of other pathophysiologic abnormalities. Further, even with extensive laboratory facilities available, there are no universally agreed upon markers which signal the impending development of overtraining syndrome. Beyond this, the length of time involved in the analysis of complex hematological or hormone variables creates a feedback loop which is too long to be of significant practical value to coaches and athletes. Certainly, at the present time, there are no simple indicators of impending overtraining syndrome that are available to coaches and athletes. Given the nearly universal tendency for coaches and athletes to respond inappropriately to temporary training or competitive incompetence by doing more

* Phone 608 785 8687, Fax: 608 785 8172, e-mail foster@mail.uwlax.edu

Overload, Performance Incompetence, and Regeneration in Sport, edited by Lehmann *et al.*
Kluwer Academic / Plenum Publishers, New York, 1999.

training, simple markers which might signal impending overtraining, or at least deteriorating overreaching, would be most useful. If one is willing to listen carefully to athletes, very often they will give cues that might signal the onset of a decompensated stage of overreaching that is the gateway to overtraining syndrome. Athletes and coaches have often monitored the resting heart rate with the thought that a persistent elevation might signal decompensation. However, since resting heart rate is affectd by so many different factors, an alternative might be to evaluate heart rate during a standard warm-up prior to training as an index of decompensation. Likewise, several investigaors have noted that athletes will consistently use certain phrases to describe their subjective response to training that seem to be predictive of overreaching which is decompensating into overtraining syndrome, including "bunt out", "flat", "washed out" or "heavy". Lehmann et al. (7, 8) have noted that a simple index of complaints appeared to track the development of overtraining syndrome in the one successful experimental model of overtraining syndrome, the increased training volume study. Others have shown that simple measures of the ratio between blood lactate and the rating of perceived exertion may signal the presence of muscle glycogen depletion or other muscular factors that may be precursors of overtraining syndrome (6, 11, 12). Given that recent technological advances have made the field evaluation of heart rate and blood lactate comparatively simple, it may be that systematic evaluation of athletes during a structured warm up would allow the coach to decide which athletes are responding inappropriately and need additional regeneration. We sought to determine whether simple laboratory measures or simple questionnaires might be of practical value in determining when athletes were entering into the decompensated stage of overreaching. Accordingly, we have adopted the strategy of evaluating athletes during a standard exercise bout near the end of their "warm up" period. We then tried to determine whether athletes who "fail their warm up" are likely to demonstrate responses consistent with the emergence of decompensated overreaching. If this strategy were to demonstrate value, then the coach might have an additional tool to manage the training program in their athletes. If valid, such an approach should be widely accepted because it would give the coach, rather than the physicain or sport scientist control of the pattern of training.

2. METHODS

The subjects of the study were competitive speed skaters, cyclists, and triathletes (n=7), ranging from national level competitors down to recreational athletes. On the average (±sd) they were 30.6±5.8 years of age, 1.82±0.07 meters in height, 80.0±4.4 kg in weight, and 11.2±4.4 % fat by skinfold measurements. Their maximal power output during cycle ergometry was 384±84 Watts or 4.79±1.01 Watts/kg, with a HRmax of 186±3 beats/min. During submaximal exercise, the power output at a blood lactate concentration of 4.0 mmol* l-1 was 251±89 Watts or 3.12±0.98 Watts/kg. The subjects were followed for 12 weeks of self-regulated training. After the first three weeks of training, the subjects were asked to generally incrase their training load. No specific direction regarding the day to day management of training was provided. During this period of time each subject performed 10km cycle time trials on a windload simulator every three weeks as a criterion measure of performance (2). The characteristics of their training program were tracked with our previously described "session RPE" method which allowed computation of the training load, training monotony and training strain (3–5). Additionally, by subtracting the most recent two weeks training load from the most recent six weeks training training load we sought to have a relative index of the fatigue which might be related to training as suggested by Bannister et al. (1, 1O). Twice weekly, each subject performed five minutes of submaximal cycle ergometer exercise as part of their pre-training warm up, at an intensity previously shown to be in the range associated with a blood lactate concentration of 4 mmol * l^{-1} (255±74 Watts). Each also completed a questionnaire intended to reveal the general level of complaints experienced by the athlete (4). At the conclusion of this exercise bout, heart rate was measured by radio telemetry, the rating of perceived exertion was measured using the category ratio scale of Borg, and blood lactate was measured in capillary blood obtained from a fingertip using an enzyme electrode system. Since this represented an exploratory study, formal statistical analysis were not performed. Instead, the data from each subject were subjected to an informal trend analysis with the intent of determining whether there was any consistent tracking between the subjects who demonstrated net favorable vs unfavorable responses to training and the responses during the standard warm-up. It was anticipated that changes in the pattern of responses during pre-training warm up would preceed changes in performance during the criterion performances.

3. RESULTS

Tracking of training loads produces understandable variations in performance. The general pattern of: a) improved training in relation to greater training loads (rolling 6 week average), and b) improved training during periods of comparative rest (defined as a positive fitness-fatigue index) was evident in most of the subjects (3). Likewise, there was a general decrease in the heart rate and percieived exertion during the standard exercise bout, with a smaller decrease in blood lactate. This was related to a slight increase in the blood lactate/RPE across the period of training. Mean (±sd) responses for the variables of interest are presented in Figure 1.

An example of training, structured warm-up and performance responses in an athlete who responded favorably to training is presented in Figures 2–3. In this subject there was steady improvement in performance, despite a generally increasing training load and a negative fitness-fatigue index. The HR during the standard exercise bout decreased steadily, although there was a "bump" during thelast week. There was a generally decreased RPE and a slight decrease in blood lactate during the standard exercise bout, with no change in the blood lactate/RPE ratio.

An example of training, structured warm-up and performances in an athlete who was responding poorly to training is presented in Figures 4–5. In this athlete there was a general stagnation of performance after day 42. This persisted despite a spontaneous decrease in training load, leading to a positive fitness-fatigue index. Perhaps because of this decrease in training load, there was no systematic increase in the complaint index. There was a general increase in HR during the warm-up bout, however, the pattern of increase in HR was not evident until after performance had already started to stagnate. There were no changes in blood lactate, RPE, or the lactate/RPE ratio.

4. DISCUSSION

Like many others before, we failed to induce any clinical evidence of serious overreaching or overtraining syndrome in this group of subjects. This observation reinforces the difficulty in producing overtraining syndrome experimentally. Although isolated cases

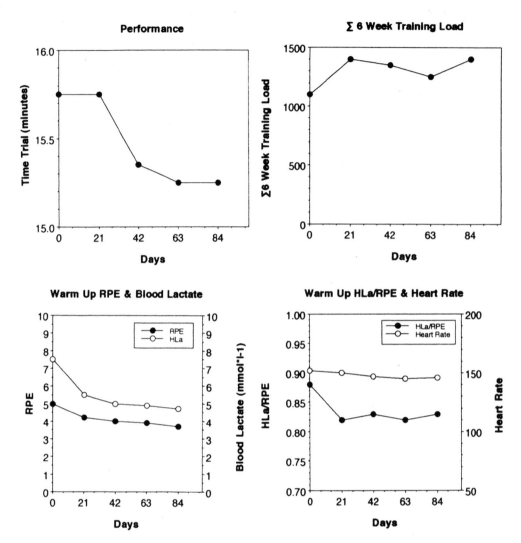

Figure 1. Mean changes in time trial performance, • six week training load, RPE, blood lactate concentration, heart rate and blood lactate/RPE ratio during the course of the training protocol. Note that there was significant improvement in performance, coupled with decreases in RPE, blood lactate and HR during a standard warm up ride; all in response to an increase in training load. There was no evidence of a systematic change in the blood lactate to RPE ratio during the warm up ride.

of overtraining syndrome have been produced in several studies, only the increased volume study (7, 8) has so far suceeded in producing overtraining in a significant percentage of subjects. There was little support for the use of the blood lactate/RPE ratio as a marker of overreaching. Although this marker had shown early promise (11), more recent data (6, 12) as well as the results of the current study suggest that it is at best a marker of transient muscle glycogen depletion. We continue to believe that the hypothesis regarding the use of a structured warm-up is is a valid concept, although support for the hypothesis is specifically lacking in the present data. It may be that only athletes training under the more

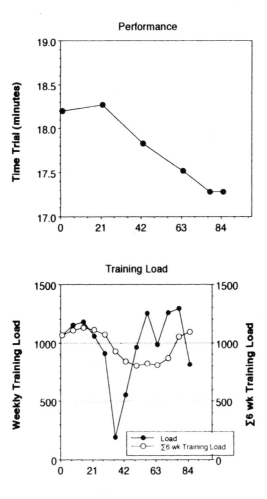

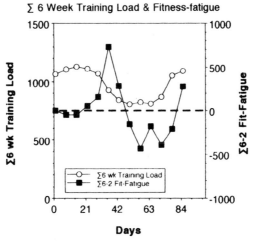

Figure 2. Changes in time trial performance (upper panel), training load and • six week training load (middle panel), and fitness minus fatigue (•6 week load - •2 week load)(lower panel) in a subject who responded well to intensified training. Note that the first large improvement in performance was associated with a decrease in training load (preceeding day 42). Note also a flattening of improvement at the end of the trial, associated with a prolonged period of relative fatigue during training (days 50–80).

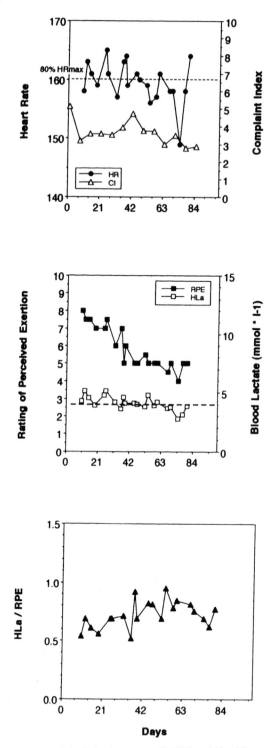

Figure 3. Changes in heart rate, complaint index (upper panel), RPE and blood lactate (middle panel), and the blood lactate to RPE ratio (lower panel) during the standard warm up ride in a subject who responded well to intensified training. Note the general decrease in heart rate and RPE, together with a smaller decrease in blood lactate, all consistent with a training response. There was no evidence of a persistently depressed blood lactate to RPE ratio that might suggest overreaching or muscle glycogen depletion.

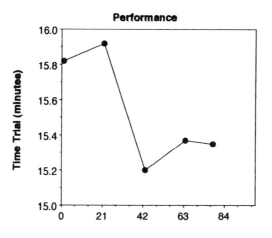

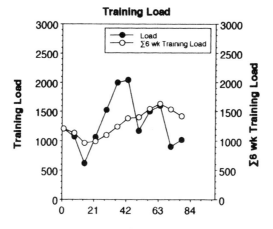

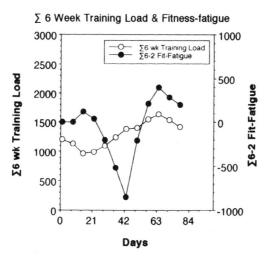

Figure 4. Changes in time trial performance (upper panel), training load and •six week training load (middle panel), and fitness minus fatigue (•6 week load - •2 week load)(lower panel) in a subject who responded relatively less well to training, with performance stagnating after day 42. Note that performance stagnated despite a decrease in the training load (with a positive fitness-fatigue index) after day 42.

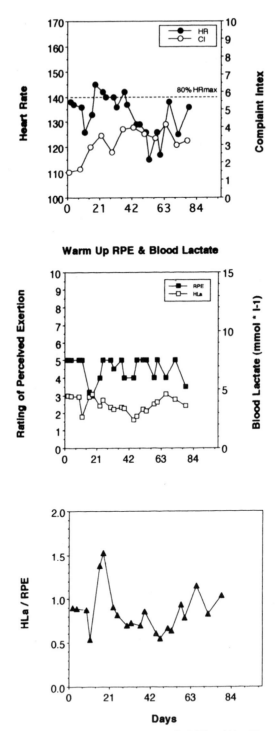

Figure 5. Changes in heart rate and complaint index (upper panel), RPE and blood lactate (middle panel) and the blood lactate/RPE ratio during the standard warm up ride in a subject who responded relatively less well to training. Note that although HR tended to decrease after day 42, there was a tendency to increase the complaint index. Also note that there was no directional change in RPE, blood lactate or the blood lactate /RPE ratio during the course of the trial.

rigorous structure of coach designed training programs, with the attendant social pressures provided by coach and teammates to ignore day to day fatigue, are likely to overreach enough to develop decompensated overreaching or overtraining syndrome. Clearly, more data collected under the circumstances of a designed training program and within the context of team preparation for serious competition is needed to provide an adequate test of this hypothesis. However, at the present time, the value of using warm-up responses as a marker of maladaptation to training remains only an interesting hypothesis

REFERENCES

1. Fitz-Clarke JAR, Morton RH, Banister EW (1991) Optimizing athletic performance by influence curves. J Appl Physiol 71: 1151–1158
2. Foster C, Green MA, Snyder AC, Thompson NN (1993) Physiological responses during simulated competition. Med Sci Sports Exerc 25: 877–882
3. Foster C, Daines E, Hector L, Snyder AC, Welsh R (1996) Athletic performance in relation to training load. Wisc Med J 95: 370–374
4. Foster C, Lehmann M (1997) Overtraining syndrome. In:Guten GN (ed) Running Injuries, Philadelphia, WB Saunders:173–188
5. Foster C (1998) Monitoring training in athletes with reference to indices of overtraining syndrome. Med Sci Sports Exerc 30: 1164–1168
6. Jeukendrup AE, Hesselink MKC, Snyder AC, Kuipers H, Keizer HA (1992) Physiological changes in male competitive cyclists after two weeks of intensified training. Int J Sports Med 13: 534–541
7. Lehmann M, Gastmann U, Petersen KG, Bachl N, Seidel A, Khalaf AN (1992) Traininging-overtraining: performance and hormone levels after a defined increase in training volume vs intensity in experienced middle distance runners. Br J Sports Med 26: 233–242
8. Lehmann M, Baumgartl P, Wiesenack C, Seidel A, Baumann H, Fischer S (1992) Training-overtraining: influence of a defined increase in training volume vs training intensity on performance, catecholamines and some metabolic parameters in experienced middle and long distance runners. Eur J Appl Physiol 64: 169–199
9. Lehmann MJ, Lormes W, Optiz-Gress A, Steinacker JM, Netzer N, Foster C, Gastmann U (1997) Training and overtraining: an overview and experimentalresults in endurance sports. J Sports Med Phys Fit 37: 7–17
10. Morton RH, Fitz-Clarke JR, Banister EW (1990) Modeling human performance in running. J Appl Physiol 69: 1171–1177
11. Snyder AC, Jeukendrup AH, Hesselink MKC, Kuipers H, Foster C: (1993) A physiological/psychological indicator of overreaching during intensive training. Int J Sports Med 14: 29–32
12. Snyder AC, Kuipers H, Cheng Bo, Servais R, Fransen E (1995) Overtraining following intensified training with normal muscle glycogen. Med Sci Sports Exerc 27: 1063–1070

5

EVALUATION OF ENDOCRINE ACTIVITIES AND HORMONAL METABOLIC CONTROL IN TRAINING AND OVERTRAINING

Atko Viru and Mehis Viru

Institute of Exercise Biology
University of Tartu
18 Ylikooli, Tartu EE2400, Estonia

1. INTRODUCTION

An essential task of hormones in metabolic control is to interference into cellular autoregulation and to ensure an extensive mobilization of resources of the body. Otherwise the actualization of potential capacities of the body is impossible. Accordingly, the exercise performance depends on influence of hormones on metabolic processes. Therefore, the magnitudes of hormonal responses in exercises, including competition performance, as well as their interrelations, allow us to understand the actual mobilization of various metabolic resources. However, in monitoring of training the significance of hormonal studies is not limited only by this approach. The determination of hormones can provide information on the adaptation to certain levels of exercise intensity and duration, as well as on disorders of adaptation, including exhaustion of the organism's adaptivity and overtraining phenomena. Hormonal responses can be used for assessment of the trainable effect of exercise session and for control of the recovery period. In order to get actually information and to avoid misunderstandings and wrong depiction, several cautions and limitations must be taken into the consideration.

2. METHODOLOGICAL CONSIDERATIONS

2.1. Body Fluids for Hormonal Determination

The most valuable information can be obtained assessing hormones in blood. The blood is the medium into which hormones are secreted by endocrine gland. With the aid of blood, hormones are transported to body, tissues. In blood, majority of hormone molecules

Overload, Performance Incompetence, and Regeneration in Sport, edited by Lehmann *et al.*
Kluwer Academic / Plenum Publishers, New York, 1999.

are bound by specific hormone-binding proteins. In most cases bound hormones are meta-bolically inactive. They cannot pass through the capillary wall and reach the interstitial compartment. It is thought that information on the actual biological effect of hormones can be obtained by free, unbound hormone fraction. However, this understanding is not complete, since a rapid exchange exists between bound and unbound fraction of a hor-mone in blood. Moreover, in stress-situation, including the exercise, increased hormone secretion leads usually to total hormone concentration that exceeds the binding capacity of plasma proteins. Accordingly, the total hormone concentration in blood rises in correlation with increase of the unbound fraction. Several methods for hormone determination in urine measure both the hormone and its metabolites together. Therefore, the hormones in urine may be used for evaluation of the general trend, but they do not provide same quan-titative characteristics as does the hormone concentration in blood. Moreover, there are at least three conditions to be accounted: (1) renal excretion of hormones depends first of all on blood level of free unbound fraction; (2) majority of molecules of hormones or their metabolites are conjugated with glucoronic or sulfuric acids before their renal excretion; therefore the value of hormone determination depends whether only unconjugated fraction or both conjugated and unconjugated fractions are determined, (3) renal excretion of hor-mones depends on kidney blood flow and rate of diuresis, both of those are reduced during exercise. The last condition makes necessary not to use hormone concentrations in urine, but hormone excretion expressed in excreted hormone amount per a time unit (usually per hour). For calculation of hormone excretion the measured concentration must be multi-plied by urine volume, excreted during the studied time period, and divided to the time of urine collection. Another approach is to use the creatinine level in urine as the reference value for evaluation of hormone concentration in urine. According to the approach urine concentrations are expressed as ratio between hormone and creatinine excretion. However, this approach cannot be appreciated in exercise studies, because the exercise-induced changes in protein metabolism can alter the creatinine production and the rate of its excre-tion. In clinics usual time period for measurement of hormone excretion is 24 h. In exer-cise studies this time period is not suitable, because it includes both the exercise and the recovery period. If there are opposite changes in hormone secretion during and after the exercise, the total 24 sample may sum up the actual increase and decrease to an un-changed hormone excretion. However, if we try to get a urine sample only for the exercise time, volume of the sample will be rather small and a methodological error is caused by low ratio between excreted urine and volume of urine remained in bladder. Another prob-lem is related to the possibility that hormones produced may not be excreted during exer-cise due to the renal retention. In conclusion, the minimal period for urine collection should be not less than 3 h. In several studies, attempts were made to assess steroid hor-mones in saliva (13, 50, 61). A great advantage of saliva sample is that frequent and easy (noninvasive) specimen collection is possible. However, only free unbound steroids are possible to enter saliva. The saliva steroids correlate with the concentration of free ster-oids in serum (13, 78). Results are reported indicating that steroid concentration in saliva is not affected by usual variation of saliva production rate (19, 78). However, this conclu-sion must be reinvestigated in regard of situations related to strong inhibition of saliva production as vigorous emotional strain or pronounced dehydration. Both situations are possible in athletes. Disadvantages of saliva hormone assessment are that the mucosa in saliva would falsely increase the saliva steroid level, and that a steroid-metabolizing en-zyme has been found in the salivary gland (67). Several hormones are found in sweat in low concentration. Their assessment does not have any purpose for training monitoring. The measurement of hormone concentration in biopsy sample of muscle tissue has meaning when in the same sample cellular hormone receptors are assessed. The last task requires to increase the amount of muscle tissue specimen. Nevertheless a source for methodological error remains due to the cellular membrane damage in biopsy sampling.

2.2. Blood Sampling

Blood for hormone analysis may be obtained from veins, arteries or capillaries. Venous blood is the specimen of choice. However, venous occlusion by the use of a tourniquet will cause fluid and low molecular weight compounds to pass through the capillary wall into the interstitial fluid. Small changes in the blood concentration of the substance occur if the tourniquet use is less than a minute, but market changes may be seen after 3 minutes. An increase of protein bound hormone by 15% after 3 min stasis has been seen. Therefore, a uniform procedure for blood collection must be used throughout an investigation, minimalizing the effect of stasis (67). Several possibilities of methodological error appear if capillary blood is collected. First, to get the necessary amount of plasma (usually 50–200 μl is required for determination of a hormone), the finger-tip or ear lobe pricks have to be sufficient to get free blood outflow. Otherwise, an additional pressure is necessary to get enough blood. When this request is satisfied, a question arises, whether the pain caused is smaller in finger-tip prick compared to venous puncture. Further, the additional pressure in order to obtain required volume of capillary blood results in two possible effects influencing the hormone concentration in sample. One is the effect of stasis, similar to action of tourniquet in venous sampling. The other is a possibility that the interstitial fluid may dilute the blood sample. An additional source of methodological error is the influence of haemolysis on hormone concentration. In whole blood the hormone concentrations are in the same range as in plasma when venous blood is analyzed. Equal hormone concentration in plasma and in whole blood is explained by the equilibrium between hormone level in plasma and erythrocytes. However, a question remains whether the equilibrium persists in exercise. Moreover, it should be taken into the consideration, that erythrocytes metabolize steroid hormones.

2.3. Conditions Influencing the Hormone Concentration in Blood

Conventional guidelines for both applied and clinical exercise testing recommend that standardized testing conditions prevail (1). Accordingly, wherever possible, researchers should ensure that exercise testing be performed at 22°C or less and a relative humidity of 60% or less, at least 2 h after individual have eaten, smoked, or ingested caffeine, at least 6 h after individual have consumed any alcohol, and at least 6 h after any previous exercise (47). Other conditions, which must be taken into the consideration and possibly standardized are nutritional status (e.g. high or low carbohydrate diet, see 22), the emotional strain (e.g. 38, 49, 73), sleep deprivation (74), circadian and seasonal rhythm, and posture. Throughout the day most blood hormones exhibit cyclic variation. The difference between maximum and minimum may be as high as 5-fold (92). A seasonal (circannual) variations in hormone levels have also been established. For example, plasma testosterone shows peak levels in the summer and a nadir in the winter (55). Hormone responses to exercise are varied to some extent in various phases of the ovarian-menstrual cycle (8). In an upright position individual's circulating blood volume is 600 to 700 ml less compared with a recumbent position. When an individual goes from supine to standing position water and filterable substances move from the intravascular space to the interstitial compartment and a reduction of about 10% in blood volume occurs. Because only protein-free fluid passes through the capillaries to the interstitial compartment, blood levels of nonfilterable substances such as protein and protein bound hormones will increase (approximately 10%). The normal decrease in blood volume from lying to standing is completed in 10 min, whereas the increase of blood volume from standing to lying is completed in approximately 30 min (67). Posture effect on blood hormone levels is also related to regulation of vascular tone: going from the supine to the standing position results in pronounced increases in levels of norepinephrine, aldosterone, angiotensin II, renin and vasopressin. In athletes an essential problem is the action of previous training regime. The hormonal studies must not be performed next day after a training session of high volume or intensity or after competition. The best time for hormonal studies is in the morning after one or two rest days. However, even in these cases one should be sure that the recovery from previous strenuous training has been completed.

2.4. Specimen Storage

After sampling a rapid separation of erythrocytes from the specimen is important since erythrocytes at room temperature can alter the plasma concentration of steroid hormones. Erythrocytes degrade estradiol to estrone and cortisol to cortisone, and absorb testosterone (see 67). Plasma has an advantage over serum since it can be removed from the erythrocytes faster and subsequently be put in cold storage. For plasma, heparin is the preferred anticoagulant over ethylendiamine tetraacetic acid (EDTA) since it cause the least interference with most tests. EDTA may cause decreases in thyrotropin, lutropin and estradiol levels in range 10 to 25%. Samples collected with EDTA yield high free testosterone results (67). Many protein hormones are thermally labile because serum or plasma samples should be stored frozen. The separation of plasma or serum should be done at 4°C. Repeated freezing and thawing of serum, plasma or urine samples should be avoided.

2.5. Hormone Analysis

The most valid method for hormone determination is radio-immuno assay (RIA). This method posses a high analytical sensitivity and specificity. In order to avoid the usage of radioactive isotopes, fluorescence immunoassay, enzyme immunoassay and chemiluminescence immunoassay are recommended. The latter is becoming the technique of choice in replacing RIA for most hormone determination. In regard of determination of monoamines (epinephrine, norepinephrine, serotonin, etc.) the high performance liquid chromatographic assay is proved to be valid.

3. INTERPRETATION OF RESULTS

3.1. Private and Attending Regulation

The hormonal regulation is actualized on two levels: (1) the level of production of signal molecules (synthesis and secretion of hormones by endocrine cells), and (2) the level of reception of signal molecules (cellular receptor proteins, which specifically bind hormones either on cellular membrane, in cytoplasm or in nucleus). Both the levels are regulated according to metabolic, homeostatic and adaptational requirements. Therefore hormone level in blood does not contain the whole information on the effects of hormonal regulation. Hormone level in blood is not the measure of hormone secretion as well, because it actually expresses the ratio between hormone inflow into the blood (secretion by endocrine gland) and outflow from the blood into the tissues (depends on dynamic balance between bound and unbound fractions, as well as on the intensity of hormone degradation in tissues).

The hormone concentration in the blood is the main determinant of the tissue supply of hormones. The quantity of hormones arriving at the tissues is divided between various sites (Figure 1). Most of hormone molecules are bound by cellular proteins. There is a dynamic equilibrium between the free hormone content in the extracellular compartment and the hormone content bound by cellular proteins. A part of the arriving hormone content is bound by sites connecting the hormone with enzymes catalyzing its metabolic degradation. This represents a loss of hormone. Beside metabolic degradation, there are also biotransformations of hormones from more active forms into less active or inactive one, or vice versa. The remainder of the active hormone content is divided into the fraction bound by specific cellular receptors of this hormone and the hormone content bound unspecifically by other proteins. The amount of hormone bound specifically by corresponding cellular receptors is

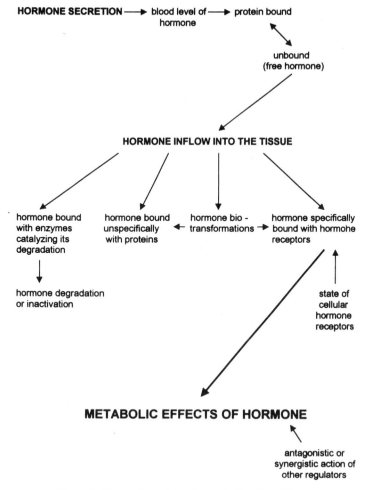

Figure 1. Factors determining the metabolic effect of hormones.

the determinant of the hormone effect. When the hormone inflow into tissues increases, more hormones can be specifically bound by their own receptors. On the other hand, when the number of hormone receptors increases an enhanced portion of hormone can be bound specifically despite the unchanged hormone content in tissues. In this connection it is valid to discriminate between 'private regulation' and 'attending regulation'(80). The 'private regulation' consists in the control of producing signal molecules (hormones). The attending regulation consists in (1) modulating influences on the number of receptors (the binding sites), on the affinity of receptor proteins to the hormone, and on the postreceptory metabolic processes; (2) regulation of metabolic situation in the cell, acting via other receptors on cellular metabolism and thus changing the actualization of private regulation, (3) regulation of protein synthesis, acting the synthesis of structure and enzyme proteins that contribute in the actualization of both private and attending regulation (Figure 2).

Other hormones may support or aggravate the actualization of the private action of a hormone. A possibility exists for blockade of the hormone action. It may take place either in a way of competition for the binding with the receptor protein between the hormone and other similar compounds or in a way of inhibiting the postreceptory metabolic processes. An example of the competition for binding with the receptor protein is the interaction between cortisol and testosterone in muscle tissue. When testosterone molecules

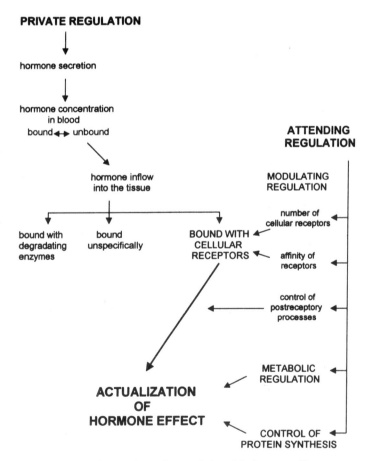

Figure 2. Private and attending regulation of the hormone effect.

occupy the specific binding sites for cortisol they exert anticatabolic effect reducing the action of cortisol on protein degradation. However, when cortisol occupies the receptors for testosterone, it exerts antianabolic action, reducing the induction of muscle protein synthesis by testosterone (46). Another example is competition for receptor between cortisol and progesterone. It has been found that glucocorticoids restore the working capacity in adrenalectomized rats inducing synthesis of regulatory protein(s). However, when together with glucocorticoid dexamethasone a great dose of progesterone was administrated, the glucocorticoid effect was inhibited, obviously, due to the ability of progesterone to be bound with glucocorticoid receptors (88). Insulin blocks the action of lipolytic hormones (epinephrine, growth hormone, glucagon) on the triglyceride hydrolysis in adipose cells. The action is related to insulin influence on postreceptory processes. Insulin activates the cAMP-phosphodiesterase and stimulates, thereby, the degradation of cAMP. The opposite, cAMP accumulation is the essential link in postreceptory processes in action of lipolytic hormones. In contrast, cortisol potentiates the epinephrine metabolic effects by changes in intracellular Ca^{2+} shifts and by inhibition of cAMP-phosphodiesterase activity (17). In conclusion, examining an increase of a hormone concentration, we cannot conclude the existence of metabolic consequences, if we do not know the interrelations with other hormones and the state of cellular hormone receptors. When the actual metabolic influence of a hormone in exercise has not checked up, the conclusions founded only on changes in the level of a single hormone are simply speculations.

3.2. Examples of Unjustified Speculations

In several cases increased cortisol level has been considered as an information on overall intensification of protein degradation, a sign of general catabolism. However, several studies provided evidence that muscular activity inhibits the catabolic action of glucocorticoids (29, 53). Therefore, in exercise the link between increased blood level of cortisol and activation of catabolic processes is not the same as in resting conditions. At the same time we cannot say that during exercise cortisol losses completely his catabolic action. Obviously, we are close to truth, if we assume, that during exercise the catabolic influence of cortisol is limited and depends on interrelation with action of several other regulators. Actually, during exercise the rate of protein degradation is elevated mainly in muscles less active (75). Experiments on rats provide evidence that exercise-induced rise in the alanine levels of blood plasma, SO fibers and liver are dependent on glucocorticoid action on the activity of alanine-aminotransferase. Adrenalectomy excluded both rise in alanine levels and increased activity of the enzyme, but substitution therapy of adrenalectomized rats with glucocorticoids restored the exercise-induced changes (86). However, infusion of testosterone reversed the exercise effect on alanine levels (23). Consequently, when exercise is performed in condition related to elevated testosterone level in blood, cortisol fails to increase alanine production. Growth hormone is known to stimulate triglyceride hydrolysis in adipose cells. However, experiments on isolated adipose cells indicated that the actualization of the growth hormone effect requires at least 2 h (18). Therefore, when anybody establishes the exercise-induced increase in growth hormone level, he can hope to have growth hormone effect only 2 h later of the onset of increased hormonal level. In accordance, during two hour cycling, the administrated exogenous human growth hormone caused a more pronounced increase in blood level of the hormone, but no increase in free fatty acid concentration in blood (66). In conclusion, suggestion about the metabolic effects of a hormone must be founded on knowledge of the time characteristics in actualization of the hormone influence on the metabolic process. Even if a correlation is found between hormonal response and metabolic change during exercise, the causal relationship can be though only if the time characteristics make real the actualization of the hormone effect. In some cases the correlation between hormone level and performance or metabolic characteristics is founded on an earlier hormone effect on muscle tissue. Recently, it was found that basal testosterone level in serum is in significant correlation with performance in counter-movement jumps and 30m dash. This fact was interpreted by causal relationship between the development of fast twitch fibers and individual differences in blood testosterone concentration (10). The effect of testosterone on fast twitch fibers development has been evidenced in pubescent animals (16). Accordingly it is possible to believe that a certain genotypic (or phenotypic) peculiarity exists, which is characterized by elevated testosterone level. The latter might promote the development of fast twitch fibers in puberty period.

3.3. Other Sources of Misevaluation of Results of Hormonal Studies

An increase of hormone concentration is not an ultimate evidence of augmented secretion of the hormone. During exercise the degradation rate of testosterone (63) and estrogen (33) decreases. It obviously contributes for increase of blood level of these hormones. Another factor, which may increase hormone level without increased secretion is the decrease of plasma volume due to extravasation of blood plasma. Already a 1-min intensive exercise bout may decrease the plasma volume by 15–20% (54). Therefore, it is desirable to calculate plasma volume changes at least by hemoglobin concentration and hematocrit value in hormonal studies. Increased hormone concentration may appear without change in secretion due to "washout" of hormones from the gland caused by increased rate of blood flow. In prolonged exercise the dehydration and rehydration may both influence the hormone levels. However, when the blood concentration of hormone increase, regardless of the mechanism, increased interaction with receptor is possible because the latter depends on actual concentration of hormone in the extracellular compartment. Each hormone response has its own dynamics. If we measure the hormone level only once during or after exercise, we may not obtain the actual picture about the response. More over, if the exercise does not require the mobilization of an endocrine function close to its maximal possibilities, individual variability may appear in pattern of hormone responses, as it has been described for cortisol (84) and ß-endorphin (90) responses (Figure 3).

The activity of endocrine systems is controlled by feedback inhibition. Therefore, the initial level of hormone may suppress or promote the hormone responses. Accordingly, from the augmented pre-exercise level the blood cortisol concentration usually drops or does not elicit any change during exercise (11, 20). Many hormones are secreted into the blood in episodic manner. The secretion bursts may be separated by rest periods with duration from 5 to 30 min (or even more). This peculiarity of endocrine functions has to be taken into the considerations in evaluation of basal hormone levels.

3.4. Determinant and Modulators of Hormonal Responses to Exercise

Variability in hormonal responses is frequently resulted by the combined influence of several factors. Misevaluation of hormonal responses is avoided if the determinants and modulators of hormone responses in exercise are accounted and used for interpretation (Figure 4).

The main determinants are exercise intensity and duration, adaptation of the person to exercise and homeostatic needs (89). It is possible to define the threshold intensity of

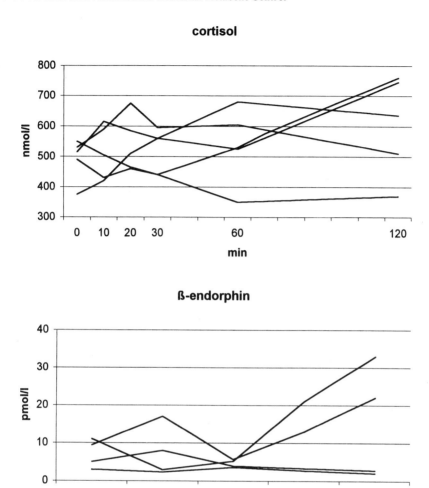

Figure 3. Individual variants of cortisol and ß-endorphin dynamics during 2-hour exercise.

exercise as the minimum intensity needed to evoke hormonal changes in blood. The dependence of the magnitude of hormonal responses on exercise duration has been also demonstrated. Therefore, the threshold duration of exercise comes also into focus. It follows the fact that exercises of underthreshold intensity may result in hormonal responses when a given amount of work has been done. At intensities above the threshold, the threshold duration is expressed by a further increase of the hormonal response or by a secondary activation of the endocrine system (Figure 5). Systematic adaptation to exercise (training) induces: (1) an increase in threshold intensity in terms of power output; thereby the training reduces, partially or totally the previously observed hormonal responses, (2) an elevated functional capacity of endocrine systems making possible very pronounced hormonal responses in extreme exercises (79, 81). The significance of exercise intensity and fitness appears first of all in activities of endocrine systems responsible for a rapid mobilization of energy reserves and protein resources. The activities of hormones controlling the water-electrolyte balance depends, first of all, on shifts in this balance.

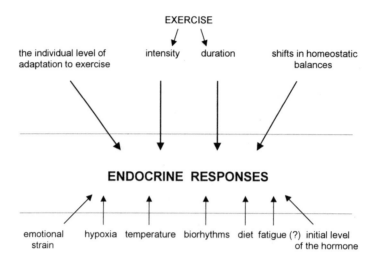

Figure 4. Main determinants and modulators of endocrine responses in exercise.

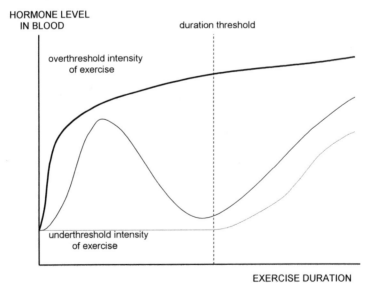

Figure 5. Effects of the duration threshold on hormone level in blood (exercise of underthreshold intensity–dotted line; exercises of overthreshold intensity–solid lines).

The actualization of the effect of main determinants is modulated by several conditions (Figure 4), among them most striking are the influence of emotional state, environmental conditions, diet (availability of carbohydrates) and biorhythms. Possible is also modulation of hormonal responses by fatigue.

4. ASSESSMENT OF THE TRAINABLE EFFECT OF A TRAINING SESSION

Training effects are founded on changes at the cellular level. In turn changes at the cellular level is the result of synthesis of structure and enzyme proteins (9) ensuring an increase of the most active cellular structures and augmentation of molecules of enzymes catalyzing the most responsible metabolic pathways. The so called "adaptive protein synthesis" is controlled at levels of induction, translation and posttranslation (9). It has been hypothesized that in the induction of the adaptive protein synthesis, the metabolites accumulating within active cells during exercise performance determine the choice of proteins, which synthesis will be intensified in postexercise period. Hormones, which secretion increases in response to the total load of training session amplify the inductory influence of metabolites (79, 82, 83). At the levels of translation the control is ensured by influence of metabolites and hormones as well (9, 79, 82). Both metabolites and hormones contribute also for adjusting the final amount of protein molecules (posttranslation control). Thus, the exercises performed influence by the accumulated metabolites the specific choice of proteins for their adaptive synthesis, while the total load on training session influences the endocrine function and thereby the contribution of hormonal control of protein synthesis. In regard of hormone inductors, most attention should be paid to changes in testosterone and thyroid hormone levels during and after training sessions for improved strength, power or endurance. The reason is that these hormones exert a strong inductory influence on synthesis of myofibrillar proteins (both hormones) and mitochondrial proteins (thyroid hormones). In postexercise period a general characteristic of the testosterone dynamics is its low level during the first hours after exercises. However, apart from endurance exercise in case of strength exercises, a tendency to the increased production of testosterone follows. This change associates with the augmentation of the testosterone and androstenedione content and increased number of androgen binding sites in skeletal muscles (65). Valuable evidence for testosterone role in training-induced hypertrophy was provided by Inoue and co-authors (31). During the training of the gastrocnemius muscles by electrical stimulation (every other day for 2 weeks), male rats were administered the androgen receptor antagonist oxendolone. The androgen receptor blockade excluded muscle hypertrophy, normally induced by electrical stimulation. In agreement with these results, significant correlation was found between the testosterone concentration and testosterone/cortisol ratio and changes in strength and power during training periods up to a year (27). Also despite the low testosterone level in women, correlation was found between serum testosterone level and the individual change in maximal force (28). This correlation might be founded on the estradiol-depending increase of sensitivity to anabolic effect of testosterone in females (15). Another possibility of compensation for the low level of testosterone is the enhanced response of somatomedin-C and other growth factors to exercise for improved strength (37). After endurance exercise the main locus of the increased rate of protein synthesis is the mitochondria of FOG and SO fibers. The highest rate of synthesis of mitochondrial and myofibrillar proteins was found 24 h after exercise. In FG fibers, the protein synthesis was suppressed within 24–48 h after an endurance exercise (87). The postexercise rise in the blood level of triiodothyronine and thyroxin coincided with the increased incorporation of ^{3}H-tyrosine in muscle fibers in normal rats. In hypothyroid rats no increase was found in label incorporation during a 48 h recovery period after 30 min running. In these rats a low level of label was found in the mitochondria as well as in all regions of sarcoplasma and myofibrils during the recovery period (36). A stimulatory effect on the protein synthesis (probably on the translational level) is produced by insulin and somatotropin. In adults with a somatotropin deficiency, the human growth hormone treatment during months increased the lean tissue and the total cross-sectional area of the thigh muscle, the strength of the hip flexors and the limb girdle muscle but not that of a number of other muscles (14). In rats daily injections of somatotropin over 36 days resulted in a significant increase in the diameter of both types of fibers (I and II) in the extensor digitorum longus and soleus muscles. The DNA/protein ratio and the number of satellite cells per muscle fiber cross-sectional area increased as well (68).

Hindleg perfusion with insulin at 200 $\mu U \cdot ml^{-1}$ stimulated protein synthesis in the white gastrocnemius. After running exercise the insulin effect was not enhanced (4).

These results together suggest that information on the trainable effect may be obtained by responses of testosterone, thyroid hormones, growth hormone and may be also of insulin. Postexercise increases in concentrations of testosterone (65), thyroid hormones (25, 36), and insulin (51) has been reported. Increase in somatotropin level has been found during the next night after exercise (2). By results of Hackney et al. (24) next night after exercise a significant augmentation of the prolactin and thyroxin responses occurred, while an attenuation of the growth hormone and cortisol responses were observed. No significant effects were found for the testosterone and lutropin pattern.

5. MONITORING OF THE ORGANISM ADAPTIVITY

Adaptivity can be defined as the ability of the organism for adequate usage of adaptation processes ensuring (1) normal life activities despite changes either the external or internal environments of body, (2) adaptive alteration in cellular structures and amount of enzyme molecules aimed to achieve a stable adaptation (resistance) to the influence of chronically acting factors. Adaptivity is the determinant of the trainable effect of repeated exercises. In athletes, the actual level of adaptivity is determined by the ratio between utilization of the adaptivity and its restoration (91).

In exercise training there is a general increase of the adaptivity. This can be explained by changes that are simultaneously essential both for increased performance and for improvement of possibilities of the mechanism of general adaptation. However, training of athletes with high load requires utilization of the organism's adaptivity to a great extent. Due to that the achieving of peak performance is related to a drop in the organism's adaptivity. Therefore, the training organization has to foresee the restoration of the organism's adaptivity. Tools for accomplishment of this task are (1) inserting training sessions with maintaining and restitution loads between sessions with trainable loads in microcycles, (2) usage of restitution microcycles after blocks of concentrated unidirectional training and after competitions, (3) relief period after competition period. The art of training consists of ensuring both purposeful utilization and restoration of the adaptivity. During a training year a significant strain of adaptation processes may be suggested in elite athletes by an increased level of blood cortisol (26, 56, 62, 76, 77). In rowers the increased cortisol resting level associated with extremely high cortisol and somatotropin concentration after a 7-min simulated rowing test (56). Increased cortisol and somatot-

ropin responses were found on the 4th, 5th and 6th microcycles of a stage of high intensive training in elite rowers (57). In swimmers 10 successive days of very intense training
resulted in an increase of serum cortisol and creatine kinase levels (35). The morning cortisol level in the blood increased during the first 4 weeks of high-intensity endurance
training and then decreased during the following 3 weeks of reduced training (64). During
7 weeks of the competition period the serum testosterone level as well as the testoster-
one/cortisol ratio decreased in rowers. A week of the regeneration phase reduced these
changes (72). Analogous changes were confirmed in other studies on endurance athletes
(26, 70). Adlercreutz et al. (3) considered the free testosterone/cortisol ratio of the plasma
to be a tool by which training could be effectively monitored. They proposed using a decrease of this ratio by more than 30% or a decrease below $0.35 \cdot 10^{-3}$ as a criterion for over-
strain, connected with extreme prevalence of catabolism. A study of elite speed skaters
confirmed the reliability of the criterion proposed by Adlercreutz and collaborators. By
the results obtained, a decrease by 30% or more in the free testosterone/cortisol ratio indicates temporary incomplete recovery from intensive training, residual weakness and, consequently, reduced effectiveness for competitive purposes (5).

A prognostic model has been elaborated by which the blood testosterone level may
be used for prognostics of the progress in strength improvement in elite weight-lifters
(12). However, the elevated cortisol level may have various functions. In the training
process, it appears not only as a result of heavy training loads. It may also appear after reduction of training loads. In young rowers (mean age 17.6 years) during training with high
volume, the serum urea content and creatine kinase activity increased, free testosterone
decreased, and cortisol remained constant. After reduction in training volume, urea and
creatine kinase normalized and free testosterone increased to a level above the initial
value. An increase was also found in the cortisol level (58).

Intensive utilization of the organism's adaptivity may be reflected also in other biochemical indices beside cortisol level. In 8 experienced middle- and long-distance runners a sharp increase in the training volume during 4 weeks caused an increased plasma
norepinephrine response to submaximal exercise (opposite to the usual training effect)
and a decrease in nocturnal excretion of epinephrine, norepinephrine, dopamine and cortisol (40, 41).

6. OVERTRAINING

After the organism's adaptivity is lost, the further training leads to overtraining. The
overtraining may appear either as an Addisonoid or as a Basedowoid syndrome (32, 39).
The parasympathetic and sympathetic types of overtraining may be also distinguished (30,
32, 39, 42, 44, 59). Some authors consider the Addisonoid syndrome and parasympathetic
type of overtraining, as well as the Basedowoid and sympathetic type of overtraining to be
synonyms (69). In addition, hypothalamic dysfunction has been described in overtraining
(6, 44). In Addisonoid form a low level of cortisol should be in case according to the ori-
gin of the term from the clinical picture in Addison disease. Actually, in older literature a
lot of data show that in exhaustive states, including overtraining, the urinary excretion of
corticosteroids and their metabolites is suppressed (for review see 81). Repeated determi-
nations of cortisol have been recommended as a tool for the diagnostics of overtraining
(44, 69). However, the behavior of blood cortisol level may be different. Some authors
found increased cortisol level at rest (6, 60). The others indicate that the exercise-induced
rise occurred to be decreased during overtraining (6, 43, 71). Accordingly, the overtraining

may be reflected in increased cortisol basal level and decreased response to exercise. Consequently, to make the blood cortisol study valuable for diagnostics of overtraining, it should be measured not only at rest but also after a certain test exercise. In accordance, impaired was also corticotrophin as well as growth hormone and insulin responses to exercise (69). In overtrained marathon runners the increase of corticotrophin, growth hormone and cortisol in response to insulin induced hypoglycemia was reduced (6). These results indicate, that if the overtraining associates with dysfunction of the adrenal cortex, the genes of this state may not be related to primary changes in adrenal cortex (remembering the Addison disease) but to pituitary or hypothalamic dysfunction. According to results of Lehmann et al. (44) in overtraining state both decreased sensitivity of adrenal cortex to corticotrophin and increased pituitary sensitivity to corticoliberin appear. The hypothalamic dysfunction is evidenced also by impaired growth hormone response to maximal exercise in overtrained athletes (69). In track and field athletes a low level of somatotropin was found when they were close to the state of overtraining (7). 22 high level middle-distance male and female runners were studied during 2 weeks of a "blow" training stage (the volume of training sessions constituted 105–120% of the usual for each athlete). Four variants of hormonal changes were found: (1) moderate increase of blood levels of cortisol and growth hormone during training sessions without any change in basal hormone levels, (2) an elevation of cortisol level in resting state together with great rises of cortisol and growth hormone concentrations during training sessions, (3) a pronounced elevation of cortisol basal level in association with a decrease of cortisol concentration during training sessions, (4) high basal level of cortisol and low levels of both cortisol and growth hormone after training sessions. The first variant was found exclusively during the first week, while the third and fourth variants were common for the end of two-weeks period (85). According to the stand-point of Adlercreutz et al. (3) overtraining must be related to decreased testosterone level, at least to a pronounced decrease in free testosterone/cortisol ratio. However in several overtraining studies essential changes were found neither in testosterone nor in cortisol levels (71). Probably the behavior of free and total testosterone as well as of cortisol is rather a physiological indicator of the current training load, but it does not necessary indicate overtraining (69).

The Basedowoid form presupposes that the levels of thyroid hormones are increased. Actually, the function of pituitary-thyroid system is not studied in aspect of overtraining. The clinical picture revealing in several overtrained athletes, remembering Basedow disease, should be better to define as sympathetic type of overtraining. This conclusion has been justified by increased blood catecholamine level (40, 42, 71). The parasympathetic type of overtraining is characterized by disturbed glycolytic energy mobilization, with reduced maximal blood lactate levels (34), bradycardia and lowered blood pressure (39, 42) and lowered plasma values of free epinephrine and norepinephrine in incremental exercise (42, 69). In the parasympathetic type of overtraining hypoglycemia has been found in athletes (39). In the sympathetic type of overtraining exercise-induced drop in insulin concentrations is exaggerated but without altered blood glucose level (71).

6.1. Conclusions

In conclusion, when decreased performance, altered cardiovascular function and changed mood prove the existence of overtraining, the hormonal studies make possible to establish the characteristics of the overtraining state and point to the probable metabolic disorders. The increased level of serotonin suggests (48) about the mechanisms of decreased performance and mood changes. In turn, the altered free tryptophan/BCAA ratio indicates the factor, promoting the serotonin action. The drop in blood glutamine level point on the mechanism of impaired immunoactivity in overtrained athletes (52). The best information about the characteristics of overtraining is provided when catecholamine, cortisol, testosterone, insulin, and growth hormone responses to test exercise are recorded.

REFERENCES

1. American College of Sports Medicine (1994) Guidelines for exercise testing and presriptcion. 4th edit. Lea & Febiger Philadelphia
2. Adamson L, Hunter WM, Ogurremi OO, Oswald I, Percy-Robb IW (1974) Growth hormone increase during sleep after daytime exercise. J Endocrinol 62:473–478
3. Adlercreutz H, Härkönen M, Kuoppasalmi K, Näveri H, Huhtaniemi TM, Tikkanen H, Remes K, Dessypris A, Karvonen J (1986) Effect of training on plasma anabolic and catabolic steroid hormones and their response during physical exercise. Int J Sports Med, 7(Suppl.1):27–28
4. Balon TW, Zorzano A, Treadway JL, Goodman MN, Ruderman NB (1990) Effects of insulin on protein synthesis and degradation in skeletal muscle after exercise. Am J Physiol 258:E92-E98
5. Banfi G, Marinelli M, Roi GS, Agape B (1993) Usefulness of free testosterone/cortisol ratio during a season of elite speed skating athletes. Int J Sports Med 14:373–379
6. Barron GL, Noakes TD, Levy W, Smith C, Millar RP (1985) Hypothalamic dysfunction in overtrained athletes. J Clin Endocrin Metab 60:803–806
7. Beyer P, Knuppen S, Zehender R, Witt D, Rieckert H, Brack C, Kruse K, Ball P (1990) Changes in spontaneous growth hormone (GH) secretion in athletes during different training period over one year. Acta Endocrin 22(Suppl.1):35
8. Bonen A, Belcastro AN (1978) Effect of exercise and training on menstrual cycle hormones. Austr J Sports Med 10:39–43
9. Booth FW, Thomason DB (1991) Molecular and cellular adaptation of muscle in response to exercise:perspectives of various models. Physiol Rev 71:541–585
10. Bosco C, Tihanyi J, Viru A (1996) Relationship between field fitness test and basal serum testosterone and cortisol levels in soccer players. Clin Physiol 16:317–322
11. Brandenberger G, Follenius M, Hietter B, Reinhardt B, Simeoni M (1982) Feedback from meal-related peaks determines diurnal changes in cortisol response to exercise. J Clin Endocrin 54:592–596
12. Busso T, Häkkinen K, Pakarinen A, Carasso C, Lacour JR, Komi PV, Kauhanen H (1990) A systems model of training responses and its relationship to hormonal responses in elite weight-lifters. Eur J Appl Physiol 61:48–54
13. Cook NJ, Ng A, Read GF, Harris B, Riad-Fahmy D (1987) Salivary cortisol for monitoring adrenal activity during marathon runs. Horm Res 25:18–23
14. Cuneo RC, Salomon F, Wills CM, Sonksen PH (1991) Growth hormone treatment in growth hormone-deficient adults. I Effects on muscle mass and strength. J Appl Physiol 70:688–694
15. Danhaive PA, Rousseau GG (1988) Evidence for sex-dependent anabolic response to androgenic steroids mediated by glucocorticoid receptors in rat. J Steroid Biochem 29:275–281
16. Dux L, Dux E, Guba F (1982) Further data on the androgenic dependence of the skeletal muscle. Horm Metab Res 14:191–194
17. Fain JN (1979) Inhibition of glucose transport in fat cells and activation of lipolysis by glucocorticoids. In: Baxter JD, Rousseau GG (eds) Glucocorticoid hormone action. Springer-Verlag Berlin, Heidelberg, New York: 547–560
18. Fain JN, Kovacov VP, Scow RO (1965) Effect of growth hormone and dexamethasone on lipolysis and metabolism in isolated fat cells of the rat. J Biol Chem 240:3522–3529
19. Fergusson DB, Price DA, Wallace S (1980) Effects of physiological variables on the concentration of cortisol in human saliva. Adv Physiol Sci 28:301–313
20. Few JD, Imms FJ, Weiner JS (1975) Pituitary-adrenal response to static exercise in man. Clin Sci Mol Med 49:201–206
21. Fry RW, Morton AR, Garcia-Webb P (1992) Biological responses to overload training in endurance sports. Eur J Appl Physiol 64:335–344
22. Galbo H (1983) Hormonal and metabolic adaptation to exercise. G.Thieme Stuttgart, New York

23. Guezennec GY, Ferre P, Serrurier B, Merino D, Amonad M, Pesquires PC (1984) Metabolic effects of testosterone during prolonged physical exercise and fasting. Eur J Appl Physiol 52:300–304
24. Hackney AC, Ness RJ, Schrieber A (1989) Effects of endurance exercise on nocturnal hormone concentrations in males. Chronobiol Intern 6:341–346
25. Hackney AC, Gulledge T (1994) Thyroid hormone responses during an 8-hour period following aerobic and anaerobic exercise. Physiol Res 43:1–5
26. Häkkinen K, Keskinen KL, Alen M, Komi PV, Kauhanen H (1989) Serum hormone concentrations during prolonged training in elite endurance trained and strength trained athletes. Eur J Appl Physiol 59:233–238
27. Häkkinen K, Pakarinen A, Alen M, Kauhanen H, Komi PV (1987) Relationships between training volume, physical performance capacity, and serum hormone concentrations during prolonged training in elite weightlifters. Int J Sports Med 8 (Suppl.1):61–65
28. Häkkinen K, Pakarinen A, Kyrölainen H, Chang S, Kim DH, Komi VP (1990) Neuromuscular adaptation and serum hormones in females during prolonged power training. Int J Sports Med 11:91–98
29. Hickson RC, Davis JR (1981) Partial prevention of glucocorticoid-induced muscle atrophy by endurance training. Am J Physiol 241:E226-E232
30. Hooper SL, Mackinnon LT (1995) Monitoring overtraining in athletes. Sports Med 1995;20:321–327
31. Inoue K, Yamasaki S, Fushiki T, Okada Y, Sugimoto E (1994) Androgen receptor antagonist suppresses exercise-induced hypertrophy of skeletal muscle. Eur J Appl Physiol 69:88–91
32. Israel S (1976) Zur Problematik der Übertrainings aus internistischer und leistungsphysiologischer Sicht. Med Sport 16:1–12
33. Keiser A, Poortmans J, Bunnik SJ (1980) Influence of physical exercise on sex hormone metabolism. J Appl Physiol 48:765–769
34. Kindermann W (1986) Overtraining–expression of a disturbed autonomic regulation. Dtsch Z Sportmed 37:238–245
35. Kirwan JP, Costill DL, Flynn MG, Mitchell JB, Fink WJ, Neufer PD, Houmard JA (1988) Physiological responses to successive days of intense training in competitive swimmers. Med Sci Sports Exerc 20:235–259
36. Konovalova G, Masso R, Ööpik V, Viru A (1997) Significance of thyroid hormones in post-exercise incorporation of amino acids into muscle fibers in rats: an autoradiographic study. Endocrin Metab 4:25–31
37. Kraemer WJ, Gordon SE, Fleck SJ, Marchitelli LJ, Mello R, Dzialos JE, Friedl K, Harman E, Maresh C, Fry AC (1991) Endogeneous anabolic hormonal and growth factor response to heavy resistance exercise in males and females. Int J Sports Med 12:228–235
38. Kreuz L, Rose R, Jennings J (1972) Suppression of plasma testosterone levels and psychological stress. Arch Gen Psychiatry 26:479–482
39. Kuipers H, Keizer HA (1988) Overtraining in elite athletes. Sports Med 6:79–92
40. Lehmann M, Baumgarte P, Wiesenack C, Seidel A, Bauman H, Fischer S, Spöri U, Gendrish G, Kaminski R, Keul J (1992) Training-overtraining:influence of a defined increase in training volume vs. training intensity on performance, catecholamines and some metabolic parameters in experienced middle- and long-distance runners. Eur J Appl Physiol 64:169–177
41. Lehmann M, Dickhuth HH, Gendrisch G, Lazar W, Thunk M, Kaminski R, Aramendi JF, Peterke E, Wieland W, Keul J (1991) Training-overtraining experimental study with experienced middle and long distance runners. Int J Sports Med 12:444–452
42. Lehmann M, Foster C, Keul J (1993) Overtraining in endurance athletes. Med Sci Sports Exerc 25:854–862
43. Lehmann M, Gastmann U, Petersen KG (1992) Training-overtraining: performance, and hormone levels, after a defined increase in training volume versus intensity in experienced middle- and long-distance runners. Brit J Sports Med 26:233–242
44. Lehmann MJ, Lormes W, Opitz-Gress A, Steinacker JM, Netzer N, Foster C, Gastmann K (1997) Training and overtraining: an overview and experimental results in endurance sports. J Sports Med Phys Fitness 1997;37:7–17
45. Lehmann M, Schnee W, Scheu R, Stockhausen W, Bachl N (1992) Decreased nocturnal catecholamine excretion: parameter for an overtraining syndrome in athletes? Int J Sports Med 13:236–242
46. Mayer M, Rosen F (1977) Interaction of glucocorticoids and androgens with skeletal muscle. Metabolism 26:937–962
47. Ministry of Fitness and Amateur Sport (1986) Canadian standardized test of fitness operations manual. 3rd edit.Canada: Minister of Supply and Services
48. Newsholme EA, Parry-Billing M, McAndrew M, Budgett R (1991) Biochemical mechanism to explain some characteristics of overtraining. In: Brouns F (ed) Medical Sports Science, vol.32, Advance in Nutrition and Top Sport. Karger Basel: 79–93

49. Pequignot JM, Peyrin L, Favier R, Flandrois R (1979) Adrenergic response to intense muscular work in sedentary man in relation to emotivity and physical training. Eur J Appl Physiol 40:117–135

50. Port K (1991) Serum and saliva cortisol responses and blood lactate accumulation during incremental exercise testing. Int J Sports Med 12:490–494

51. Pruett EDR (1985) Insulin and exercise in non-diabetic and diabetic man. In: Fortherby K, Pal SB (eds) Exercise Endocrinology. W.de Gruyter Berlin, New York: 1–23

52. Rowbotton DG, Keast D, Morton AR (1996) The emergency role of glutamine as an indicator of exercise stress and overtraining. Sports Med 21:80–97

53. Seene T, Viru A (1982) The catabolic effect of glucocorticoid on different types of skeletal muscle fibers and its dependence upon muscle activity and interaction with anabolic steroids. J Steroid Biochem 16:349–352

54. Sejersted OM, Vollestand NK, Melbo JI (1986) Muscle fluid and electrolyte balance during and following exercise. Acta Physiol Scand 128(Suppl 556):119–127

55. Smals AGH, Kloppenborg PWC, Benraad TJ (1976) Circannual cycle in plasma testosterone levels in man. J Clin Endocrin Metab 42:979–982

56. Snegovskaya V, Viru A (1993) Elevation of cortisol and somatotropin levels in the course of further improvement of performance capacity in trained rowers. Int J Sports Med 14:202–206

57. Snegovskaya V, Viru A (1992) Growth hormone, cortisol and progesterone levels in rowers during a period of high intensity rowing. Biol Sport 9:93–102

58. Steinacker JM, Laske R, Hetzel WD, Lormes W, Liu Y, Stanch M (1993) Metabolic and hormonal reactions during training in junior oarsmen. Int J Sports Med 14(Suppl. 1): S24-S28

59. Stone MH, Keith RE, Kearney JT, Fleck SJ, Wilson GD, Triplett NT (1991) Overtraining: a review of the signs, symptoms and possible causes. J Appl Sport Sci Res 5:35–50

60. Stray-Gundersen J, Videman T, Snell PG (1986) Changes in selected objective parameters during overtraining. Med Sci Sports Exerc 18:S54-S55

61. Stupnicki R, Obminski Z (1992) Glucocorticoid responses to exercise as measured by serum and salivary cortisol. Eur J Appl Physiol 65:546–549

62. Stupnicki R, Obuchowicz-Fidelus B, Jedlikowski P, Kuslewicz A (1992) Serum cortisol, growth hormone and physiological responses to laboratory exercise in male and female rowers. Biol Sport 9:17–23

63. Sutton JR, Coleman MJ, Casey JH (1978) Testosterone production rate during exercise. In: Landry F, Orban WA (eds) 3rd International Symposium on Biochemistry of Exercise. Symposia Specialists Miami: 227–234

64. Tabata I, Atomi Y, Misyashita M (1989) Biphasic change of serum cortisol concentration in the morning during high-intensity physical training in man. Horm Metab Res 21:218–219

65. Tchaikovsky VS, Astratenkova IV, Bashirina OB (1986) The effect of exercise on the content and receptor of the steroid hormones in rat skeletal muscle. J Steroid Biochem 24:251–253

66. Toode K, Smirnova T, Tendegolskis Z, Viru A (1993) Growth hormone action on blood glucose, lipids and insulin during exercise. Biol Sports 10:99–105

67. Tremblay MS, Chu SY, Mureika R (1995) Methodological and statistical considerations for exercise-related hormone evaluation. Sports Med 20:90–108

68. Ullman M, Oldfors A (1986) Effects of growth hormone on skeletal muscle. I Studies on normal adult rats. Acta Physiol Scand 135:531–536

69. Urhausen A, Gabriel H, Kindermann W (1995) Blood hormones as markers training stress and overtraining. Sports Med 20:351–376

70. Urhausen A, Kindermann W (1992) Biochemical monitoring of training. Clin J Sports Med 2:52–61

71. Urhausen A, Kindermann W (1994) Monitoring of training by determination of hormone concentration in the blood–review and perspectives. In: Liesen H, Weiß M, Baum M (eds) Regulations und Repairmechanismen. Deutscher Ärzte-Verlag Köln: 551–554

72. Urhausen A, Kullmer T, Kindermann W (1987) A 7-week follow-up study of the behaviour of testosterone and cortisol during the competition period in rowers. Eur J Appl Physiol 56:528–533

73. Vaernes R, Ursin H, Darragh A (1982) Endocrine response patterns and psychological correlates. J Pychosom Res 26:123–131

74. Vanhelder T, Radomski MW (1989) Sleep deprivation and the effect on exercise performance. Sports Med 7:235–247

75. Varrik E, Viru A, Ööpik V, Viru M (1992) Exercise-induced catabolic responses in various muscle fibers. Can J Sports Sci 17:125–128

76. Vervoorn C, Quist L, Vermulst L, De Vries W, Thijssen HH (1991) The behaviour of the plasma free testosterone/cortisol ratio during a season of elite rowing training. Int J Sports Med 12:257–263

77. Vervoorn C, Vermulst L, Koppenschaar HPE, Erich WBM (1992) Seasonal changes in performance and free testosterone: cortisol ratio of elite female rowers. Eur J Appl Physiol 64:14–21
78. Vining RF, McGinley RA, Makarytis JJ, Ho KY (1983) Salivary cortisol: A better measure of adrenal cortical function than serum cortisol. Ann Clin Biochem 20:329–335
79. Viru A (1995) Adaptation in Sports Training. CRC Press Boca Ration, Ann Arbor, London, Tokyo
80. Viru A (1991) Adaptive regulation of hormone interaction with receptor. Exp Clin Endocrin 97:13–28
81. Viru A (1985) Hormones in Muscular Activity. Vol.2. Adaptive Effects of Hormones in Exercise. CRC Press Boca Raton Fl
82. Viru A (1994) Molecular cellular mechanisms of training effects. J Sports Med Fitness 34:309–322
83. Viru A (1984) The mechanisms of training effects: A hypothesis. Int J Sports Med 5:219–227
84. Viru A, Karelson K, Smirnova T (1992) Stability and variability in hormone responses to prolonged exercise. Int J Sports Med 13:230–235
85. Viru A, Kostina L, Zhurkina L (1988) Dynamics of cortisol and somatotropin contents in blood of male and female sportsmen during their intensive training. Fiziol zhurn (Kiev), 34(4):61–66
86. Viru A, Litvinova L, Viru M, Smirnova T (1994) Glucocorticoids in metabolic control during exercise: alanine metabolism. J Appl Physiol 76:801–805
87. Viru A, Ööpik V (1989) Anabolic and catabolic responses to training. In: Kvist M (ed) Paavo Nurmi Congress Book. The Finnish Society of Sports Medicine Turku: 55–56
88. Viru A, Smirnova T (1985) Involvement of protein synthesis in the action of glucocorticoids on the working capacity of adrenalectomized rats. Int J Sports Med 6:225–228
89. Viru A, Smirnova T, Karelson K, Snegovskaya V, Viru M (1996) Determinants and modulators of hormonal responses to exercise. Biol Sport 13:169–187
90. Viru A, Tendzegolskis Z, Smirnova T (1990) Changes of ß-endorphin level in blood during prolonged exercise. Endocrinol exp 24:63–68
91. Viru A, Viru M (1997) Adaptivity changes in athletes. Coaching and Sport Sci J 2(2):26–35
92. Weitzman ED (1976) Circadian rhythms and episodic hormone secretion. Ann Rev Med 27:225–243

6

CLINICAL FINDINGS AND PARAMETERS OF STRESS AND REGENERATION IN ROWERS BEFORE WORLD CHAMPIONSHIPS

J. M. Steinacker,[1] M. Kellmann,[4] B. O. Böhm,[3] Y. Liu,[1] A. Opitz-Gress,[1] K. W. Kallus,[5] M. Lehmann,[1] D. Altenburg,[2] and W. Lormes[1]

[1]Department of Sports Medicine
[3]Department of Internal Medicine I, Endocrinology
Medical Hospital
University of Ulm
89070 Ulm, Germany
[2]German Rowing Association
30189 Hannover, Germany
[4]Institute for Sports Science
University of Potsdam
Am Neuen Palais 10
14469 Potsdam, Germany
[5]Institute for Psychology
Dep. of Applied Psychology
Karls-Franzens-University Graz
Universitätsplatz 2
A-8010 Graz, Austria

1. INTRODUCTION

The evaluation of the clinical state of an athlete, e.g. of current trainability and of the diagnosis of overload and overtraining, is already one of the most complicated tasks in athletic medicine (8, 21, 23, 24, 26, 40). Training is not only repetitive physical exercises, but also regular regeneration as an integral part of a successful training program (1, 7, 26, 38). As already shown in several experimental studies, clinical, metabolic and hormonal findings, including the psychologically-related monitoring of stress and recovery, seem to reflect the clinical state of athletes (5, 18, 19, 20, 21). Such parameters can be used to monitor training and regeneration in athletes. However, there remains some uncertainty concerning reliability of such parameters for monitoring training and regeneration of elite

Overload, Performance Incompetence, and Regeneration in Sport, edited by Lehmann *et al.*
Kluwer Academic / Plenum Publishers, New York, 1999.

athletes during periods preparatory to major events like World Championships (1, 11, 30, 36, 40). This review deals with some of the aspects of these important practical and scientifical questions based on the experience of several preparatory training camps in rowing. Rowing has to be seen as strenuous middle time endurance stress of 5.5 to 8.0 minutes' duration, for which the athletes perform hughe training programs in which the monitoring of adaptation is essential to prevent long-term overtraining.

2. PERFORMANCE

Besides mental stability and rowing technique, improvement of physical performance is the specific goal of training. Therfore, performance is the most important parameter for monitoring training adapation. Changes in performance capacity can be analyzed during "all-out" rowing tests in the boat over various distances or during rowing ergometer tests. Maximum performance (Pmax) during a standardized test can be used for evaluation of the exercise capacity (10, 26, 35, 36, 40). However, Pmax is subject to motivation of the rower tested and thus may not be sensitive enough for monitoring a complete rowing season (38). Therefore, more reliable test programs such as fast ramp tests or short maximum tests are under discussion. Such underdistance tests may be helpful because maximum tests may be not fit into a training schedule at any time. Boat velocity is another very important parameter for performance. Although it is influenced by the surroundings, it can be be also measured during shorter than competion distance.

Table 1 presents the time course of rowing velocity and ergometer results for a coxed eight during the training camp before Junior World Championships 1995 (41). Such a preparatory training program typically has a duration of approximately 4 weeks: 2 weeks high-intensity / high-volume training, tapering 1 week and preparation for the finals in the last week. Rowing speed for the 2000 m distance was slowest during the high-volume / high-intensity phase 2, and the fastest boat speed was observed at time trial 4 after the tapering period (phase 3) and at the World Championships.

Similar results are reported in other studies. For example, a decrease in performance was observed in recreational athletes after 7 to 10 successive days of unaccustomed prolonged training (5, 6, 9). In swimmers, performance decreased after 10 days of intensified swimming training (4), and in cyclists, after 14 days of intensive training (14). Decreased

Table 1. Performance 1) in the coxed eight (2000 m time; t_{2000m}) at the National Championships, at the time trials TT1–TT4, and at the final of the World Championship, and 2) in the incremental rowing ergometer tests 1 and 2: maximum power (P_{max}), maximum lactate concentration (La_{max}) and Power at 4 $mmol \cdot L^{-1}$ lactate (P_{LAT})[a]

	Nch	TT1	TT2	TT3	TT4	WhC
$t_{2000\,m}$ [min:sec]	6:04.15	6:12.36	–	6:07.27	5:44.31	5:51.23
		RE1			RE2	
P_{max} [Watt]		465.8 ± 20.8		+	485.6 ± 16.8	
La_{max} [mmol.L-1]		16.5 ± 2.8			15.8 ± 3.1	
P_{LAT} [Watt]		350.0 ± 20.7		+	369.1 ± 30.4	

n= 10, median (minimum - maximum)

+: p< 0.05 between RE1 and RE2

[a]Modified from Steinacker et al. (41).

performance was followed by supercompensation after 1–2 weeks of recovery (4, 14). Increases in training volume are generally considered to be more critical than increases in training intensity at moderate total training load (9, 10, 23, 25).

The maximum oxygen uptake (VO2max) has the best predictive value for competetive results (at a given rowing efficiency, 11, 35). A large body muscle mass is involved and VO2max reaches in world class oarsmen 6.0 to 6.7 $L\,min^{-1}$ (65 to 75 $mL\,min^{-1}kg^{-1}$ (11, 30, 36, 37, 38). The VO2max increases with distance rowed, but levels off at training volumes of approximately 5000 to 6000 km per year (36, 38). Seasonal changes of VO2max have been described from 5 to 20 $mL\,min^{-1}kg^{-1}$ during a competition season (11, 36, 38). VO2max decreases in highly-trained athletes when the weekly distance rowed is reduced below approximately 100 $km\,week^{-1}$ (36, 38). Despite the significance of VO2max, the measurement is costly, time consuming and needs laboratory and personnel capacities.

Endurance capacity is an important result of training and regeneration (30, 40). Higher performance at a fixed or individual lactic threshold means higher maximum performance, but there is a wide scattering of individual data. In successful rowers, the 4 $mmol\,L^{-1}$ lactate threshold is in the range of 75 to 85 % of Pmax (38, 40). With a higher percentage of type I fibers, more power per stroke can be performed at a given blood lactate concentration of 4 $mmol\,L^{-1}$ (11, 35, 38). Specific endurance training increases the work per stroke at the same lactate level; this may be based on muscular type I fiber content but also on increased oxidative capacity of type II fibers (35, 38). After exhaustive rowing, maximal blood lactate concentration is lower when the lactic threshold is higher, which means higher oxidative and lower glycolytic capacities. However, lower muscular glycolytic capacities may negatively influence start and final spurt during a rowing race (35, 36). Although lactic threshold and maximum lactate are influenced by preceding exercise and muscle glycogen stores and thus also the interpretation of lactate-performance curves, their importance for the analysis of sport-specific endurance capacity is accepted (28, 29, 35, 38). Therefore, all variables before a test have to be standardized and depletion of glycogen stores avoided using nutrition with sufficient carbohydrate content. In a glycogen- deficient state, maximum lactate and performance are depressed and lactic threshold virtually increased. In contrast, in the state of overreaching or overtraining without glycogen deficit, maximum lactate and performance capacity and lactic threshold are decreased or lactic threshold is unchanged (23, 25, 26, 40).

In a an initial conclusion, training and overload situations and the effect of regeneration can be monitored using laboratory or field tests, even covering underdistances.

3. CLINICAL FINDINGS

The rate of various physical complaints rises with increased training load. It is under discussion, whether high training loads will lead to a higher rate of infections (open window theory, Pedersen (33)). In a 5-week training camp of the junior rowing team (1996), the male rowers showed a clear peak in visits to the doctor in the period with the highest training load (Table 2) in aggreement with previous studies which report an increase of symptoms with increased training load (8, 13, 26). However, in the phase with increased number of visits to the doctor, clearly no overload or overreaching situation was attained (see discussion below).

The percentage of visits due to bacterial or viral infections varied between 12 and 47 %, and the percentage of visits due to muscular or orthopedic problems was between 10

Table 2. Total number of visits of the male rowers to the doctor
in the training period before the Junior World
Championships 1996 (n=30)

Training phase	1	2	3	4	5
Total	8	28	13	19	12
Infection	1	11	4	9	5
Muscle / Orthopedic	3	3	5	4	5
Other	4	11	4	7	1

and 41 % of all visits to the doctor. Acute complaints or illnesses are not a specific sign of
overload. They may also be related to other, not training-dependent factors like exposure
to virulent agents or accidents and they are therefore often nonspecific. However, in single
subjects, we clinically observed infections, like herpes zoster, which are clearly due to depressed immunity (frequency approximately $1 \cdot 500^{-1} \cdot$athletes$^{-1} \cdot$week^{-1}). On the other hand,
there is evidence that during the preparation of rowers for championships, immune func-
tion is normal in most athletes. Thus, Jakeman et al. (13) observed changes in neopterin
excretion as a marker of the activation of the cellular immune system in 27 oarsmen before the Olympics. They found an increase in neopterin excretion consistent with mono-
cyte/macrophage activation (Figure 1). However, training involves adaptation or
destruction or damage of skeletal muscle fibers, and monocyte/macrophage activation
may be related to tissue damage, stress and decay of fibers, which may trigger an increase
in muscular heat shock protein synthesis in rowers during heavy training (27), resulting in
an increased muscular resistance to stress.

4. SELECTED BIOCHEMICAL AND HORMONAL INDICES OF OVERREACHING

Biochemical values are often used to monitor training load. Serum creatin kinase activity increases at the beginning of a training cycle primarily reflecting muscular load,

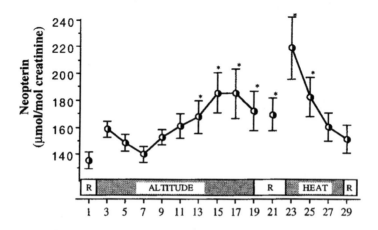

Figure 1. Changes in urinary neopterin excretion of 27 elite oarsmen during training in preparation for Olympic
competition. Means and standard deviation. From Jakeman et al. (13) with permission.

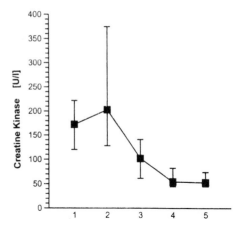

Figure 2. Creatin kinase during the training camp of the German Junior National Team 1996, at the beginning (1), and after training phases of approximately one week (2–5). Phase 2 and 3 are high volume training phases, phase 4 is tapering and phase 5 is the training before the heats of the World Championships. Median values and the interquartile range (25–75%).

whereas normalization of CK activity during training reflects the muscular adaptation during that training cycle (29) (Figure 2). Serum uric acid concentration can be used as an indicator of alactic anaerobic load due to accumulation of AMP, it increases with weight training and sprint training in rowing (40). But both parameters are less specific to detect an overreaching or overtraining situation and are rather used for analysis of the actual training load.

Hormonal responses were often proposed for monitoring overreaching and overtraining situations; however, of the many hormonal values, only serum cortisol and basal catecholamine excretion currently have practical value and are discussed in this review (3, 6, 12, 21, 23, 44, 45). Examinations of the response of the hypothalamo-pituitary-adrenocortical axis have increased the knowledge about hormonal regulation in training, overreaching, overtraining and regeneration (3, 10, 24, 43). It was already shown that basal follicle stimulating hormone (FSH) was depressed during high-load training phases of rowers suggesting hypothalamic downregulation (41). This is in accordance with a hypothalamic hyporesponsiveness as decribed by Lehmann et al. and Barron et al., who observed that insulin-dependent hypoglycemia or corticotropin-releasing hormone (CRF) went along with a lower adrenocorticotropin (ACTH) or / and cortisol response in the state of overtraining (2, 24, 26). During high-load training periods, peripheral and central steroid hormone responses decrease, and during tapering, normalize again (21, 25, 26, 41, 43).

A downregulation of hypothalamo-pituitary-adrenocortical axis will result in lower basal cortisol levels (24, 26, 42, 43). Therefore, basal cortisol levels may be used for analysis of the effects of training. At an early stage during the training camp 1996, when training load was highest, basal cortisol levels increased by 18 % and decreased slightly afterwards (Figure 3). This can be indicative of a metabolic problem (e.g. glycogen deficiency) or of increased training-dependent stress (22, 26, 42). A common metabolic cause can be seen in glycogen depletion with increased counterregulatory activity of hormones such as cortisol, growth hormone and catecholamines (2, 24, 42). Elevated basal cortisol levels are often seen as a normal stress response to high-intensity training (39, 40, 42), whereas decreased basal cortisol levels and decreased pituitary-adrenocortical responsiveness are a late sign of overreaching or overtraining (10, 26). Comparing the training of the rowing teams 1995 and 1996, cortisol levels were 20 % lower (450 nmol·L^{-1}) during the first weeks and increased during tapering by 10% in 1995 (1989: 23 %) (39, 41). During the 1996 camp, the decrease of cortisol during tapering was not as pronounced as in 1995,

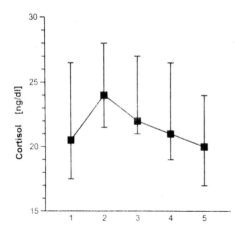

Figure 3. Resting morning cortisol concentrations at the beginning of a training camp 1996 (1), and after the training phases (2–5), 5 is before the World Championships. Median values and the interquartile range (25–75%). Further explanation see Figure 2.

suggesting that overreaching was only moderate. Furthermore it can be speculated that training load in the weeks before the training camp was lower in 1996 than 1995, or, there might have been additional metabolic problems, and/or regeneration could have been insufficient. As discussed above, basal cortisol levels reflect an endpoint of the hypothalamic-pituitary-adrenocortical axis. Testing the axis by functional tests will give much more information, but they are time- and cost-consuming and are stressful for the athletes (3, 10, 24, 43). Therefore, such tests can only be performed during controlled studies.

The excretion of free catecholamines during overnight rest can be seen as the basal renal (urinary) catecholamine excretion reflecting the intrinsic activity or tone of the sympathetic nervous system, as activating mechanisms are clearly reduced during night rest (22, 42). Because noradrenaline concentrations in plasma and in cerebrospinal fluid are quite similar, circulating and excreted noradrenaline may also reflect the neuronal noradrenaline release in the brain. Downregulation of the intrinsic sympathetic nervous system activity has been demonstrated in studies by Naessens (32) and Lehmann (22) as a late finding in overtrained athletes. Resting and submaximal plasma noradrenaline concentrations can be increased in this stage (9, 12, 22, 26) going along with a decreased ß-adrenoreceptor density as described by Jost (15) during high-volume training in swimmers and runners. During the 1996 camp, an increase in basal catecholamine excretion was observed during training which is consistent with an adaptation to increased training stress (26, 42) in accordance to the cortisol data, thus excluding profound overreaching / overtraining in most of the athletes.

5. MOOD STATE AND PSYCHOLOGICAL MONITORING OF TRAINING LOAD

Monitoring of the athlete's mood state has also been introduced to rowing training (19, 20, 34). The "Recovery-Stress-Questionnaire for Athletes" (RESTQ-Sport) can be used to quantitatively measure the actual level of current stress imposed on athletes, taking regeneration scores into consideration additionally. Studies of German and American elite athletes have demonstrated that the 19 scores of the RESTQ-Sport in German and in English have acceptable levels of reliability (16, 17, 18). Moreover, previous longitudinal studies have shown that the RESTQ-Sport can sensitively monitor stress and recovery processes in training camps and throughout a total season (18, 19, 41). Thus, a dose-response relationship was demonstrated between training load, performance ability and somatic components of stress and recovery in rowers (19, 41).

This approach is somewhat different from that recommended by Morgan et al. and Hooper et al. (12, 31, 34) which monitored symptoms associated with overtraining and staleness, thus demonstrating the positive modification by success of disturbances in mood state related to training load in female rowers during a training season (31, 34) and marked mood disturbance in the state of overtraining (12, 26). Monitoring the current levels both of stress and recovery has the possible advantage that problems may be detected before symptoms of overtraining and staleness (e.g., drowsiness, apathy, fatigue, irritability) are likely to appear (17, 18, 20). However, stress and recovery are often different in their time course. Figure 4 demonstrates data from 1996. The values of the score "Somatic Complaints" increased early and decreased afterwards with adaptation. The score for "Fatigue" rose from low levels during the first 2 weeks to the third time point, with a decrease then and at the lowest approaching the World Championships. In contrast to the scores for "fatigue" and "Somatic Complaints", the scores of "General Stress" were low and quite stable during the training camp, with a small increase at the beginning and a decrease before the preliminary heats.

The changes in the recovery scales were also not uniform. Early there are no relevant changes except in the "Somatic Relaxation" score which fits somewhat to the stress-score "Somatic Complaints". The scores of "General Wellbeing" and "Social Relaxation" show lowest values at the fourth measurement and a small increase at time point 5, however, basal values are not attained again. In general, recovery can be considered not to be adequate in this training camp. However, all results of RESTQ-Sport presented here were available to coaches as individual profiles of the recovery-stress state (19, 20) and training may have been modified accordingly.

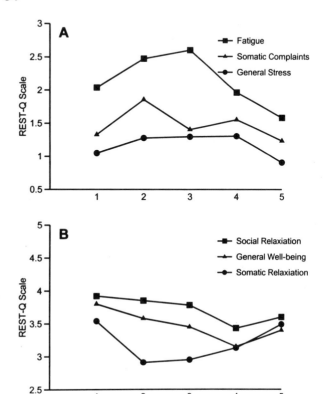

Figure 4. Results of the Recovery-Stress-Questionnaire for Athletes (RESTQ-Sport) at the beginning of a training camp 1996 (1), and after the training phases (2–4), 5 is before the World Championships: Parameters of stress (scales 'fatigue', 'Somatic Complaints', 'General Stress') (upper panel). Parameters of recovery (scales 'Social Relaxation', 'General Wellbeing', 'Somatic Relaxation') (lower panel). Mean values, n=20. Further explanation see Figure 2.

The interpretation of the indices of stress and recovery before the finals indicated low levels of stress but incomplete regeneration. This is in accordance with the hormone data.

6. APPROACH TO AN INTEGRATIVE ANALYSIS OF TRAINING

There is a cascade of various responses to prolonged training, which can be used to monitor an athlete. It is also evident that only a few parameters are reliable and specific enough. One should also distinguish between parameters in which the individual response is different between different subjects, like CK activity, hormonal parameters or mood scales, and parameters which are "hard", like physical performance.

Performance is the definite key parameter for analysis of the adaptive state of an athlete to a training program. Performance incompetence is the key symptom of overreaching or overtraining and can be specifically tested using sports-specific performance tests (21, 22, 26, 40). This can be done by rowing ergometer tests, by field tests, or, retrospectively, by comparing competition results or medal ranking. Comparing the 1995 training program (40) and the data presented here, the average medal ranking was higher in 1995 (5 medals, ranking 21 (best rank for 8 male boats is 8, which means 8 times gold) compared to 1996 (4 medals, ranking 37). During ergometric testing, maximum power of all male athletes amounted to 476.2 ± 27.0 Watt and 459.3 ± 25.7 Watt, 1995 and 1996, respectively. These different values are reflected by additional parameters such as mood state, or hormonal responses, as discussed above.

This may indicative, that overreaching is an integral part of a successful training program (1, 22, 25, 40).

For monitoring, it is also important that mood is correlated to physical performance ability, hormonal parameters and metabolic data. Thus, a U-shaped relation between subjective fatigue-ratings and sympathetic tone, e.g. basal noradrenaline excretion, was demonstrated by Naessens (32). Mood state can be monitored using the RESTQ-Sport, but different subscales have a different time course (19, 20). Somatic complaints are highest with highest training load elevated cortisol concentrations and high CK activity, but are lower when fatigue score increases. Fatigue peaks together with sympathetic activation (noradrenaline secretion). Therefore, stress and recovery in athletes have to be analyzed in a multivariate approach.

7. CONCLUSION

The diagnosis of training adaptation and the clinical state of an athlete is a complex task. However, the underlying mechanisms are more and more understood, so that specific diagnostic tools can be used sufficiently for monitoring training. CK activity and uric acid concentrations are indicative of the actual training load and less of an overtraining state. Basically, overreaching and overtraining mean performance incompetence, and performance incompetence has to be diagnosed by performance tests. Mood state can be reliably measured by the Recovery-Stress-Questionnaire for Athletes and presents further information concerning the adaptation state of athletes. Beside other hormonal parameters, cortisol serum levels may be important for the analysis of the actual metabolic and nonmetabolic stress as the endpoint of the hypothalamic-pituitary-adrenocortical axis. Additionally, nocturnal urinary noradrenaline excretion may be a simple tool to evaluate the tone of the sympathoadrenergic system which is related to adaptation/dysadaptation of athletes.

ACKNOWLEDGMENTS

The study was supported by grants from the Bundesinstitut für Sportwissenschaft, Cologne. We thank the coaches and participating athletes of the team for their close cooperation and the German Rowing Association (Deutscher Ruderverband), especially Jürgen Dabrat and Michael Müller, for substantial support. Helga Bach, Cornelia Bauer and Kristin Vilupek were excellent technical assistants to the team.

REFERENCES

1. Bannister EW, Morton RH, Clarke JR (1997) Clinical dose-response effects of exercise. In: Steinacker JM, Ward SA (eds) The physiology and pathophysiology of exercise tolerance. Plenum London New York: 297–309

2. Barron JL, Noakes TD, Lewy W, Smith C, Millar RP. Hypothalamic dysfunction in overtrained athletes (1985) J Clin Endocrinol Metabol 60:803–6

3. Bruin D (1994) Adaptation and overtraining in horses subjected to increasing training loads. J Appl Physiol 76:1908–13

4. Costill DL, Flynn MG, Kirwan JP, Houmard JA, Mitchell JB, Thomas R, Sung HP (1996) Effects of repeated days of intensified training on muscle glycogen and swimming performance. Med Sci Sports Exerc 20:249–254

5. Dressendorfer RH, Wade CE, Claybaugh J, Cucinell SA, Timmis GC (1991) Effects of 7 successive days of unaccustomed prolonged exercise on aerobic performance and tissue damage in fitness joggers. Int J Sports Med 12:55–61

6. Dressendorfer RH, Wade CE (1991) Effects of a 15-d race on plasma steroid levels and leg muscle fitness in runners. Med Sci Sports Exerc 23: 954–6

7. Foster C, Hector LI, Welsh R, Schrager M, Green MA, Snyder AC (1995) Effects of specific vs cross training on running performance. Eur J Appl Physiol 70:367–372

8. Foster C (1998) Monitoring training in athletes with reference to overtraining syndrome. Med Sci Sports Exerc 30: 1164–1168

9. Fry RW, Morton AR, Garcia-Webb P, Crawford GPM, Keast D (1992) Biological responses to overload training in endurance sports. Eur J Appl Physiol 64:335–344

10. Hackney AC, Sinning WE, Bruor BC (1990) Hypothalamic-pituitary-testicular axis function in endurance-trained males. Int J Sports Med 11:298–303

11. Hagerman FC, Staron RS (1983) Seasonal variations among physiological variables in elite oarsmen. Can J Appl Spt Sci 8:143–148

12. Hooper SL, Mackinnon LT, Howard A, Gordon RD, Bachmann AW (1995) Markers for monitoring overtraining and recovery. Med Sci Sports Exerc 27: 106–112

13. Jakeman PM, Weller A, Warrington G (1995) Cellular immune activity in response to increased training of elite oarsmen prior to Olympic competition. J Sports Sci 13: 207–211

14. Jeukendrup AE, Hesselink MKC, Snyder AC, Kuipers H, Keizer HA (1992) Physiological changes in male competitive cyclists after two weeks of intensified training. Int J Sports Med 13:534–541

15. Jost J, Weiss M, Weicker H (1990) Sympathoadrenergic regulation and the adrenoceptor system. J Appl Physiol 68:897–904

16. Kallus KW (1995) Der Erholungs-Belastungs-Fragebogen [The Recovery-Stress-Questionnaire]. Swets & Zeitlinger Frankfurt

17. Kallus KW, Kellmann M (1998) Burnout in sports: a recovery-stress state perspective. In: Hanin Y (Ed.) Emotion in Sport. Human Kinetics Champaign: in press

18. Kellmann M (1997) Die Wettkampfpause als integraler Bestandteil der Leistungsoptimierung im Sport: Eine empirische psychologische Analyse [The Rest Period as an Integral Part for Optimizing Performance in Sports: An Empirical Psychological Analysis]. Dr.Kovac Hamburg

19. Kellmann M, Kallus KW, Günther K-D, Lormes W, Steinacker JM (1997) Psychologische Betreuung der Junioren-Nationalmannschaft des Deutschen Ruderverbandes [Psychological consultation of the German Junior Rowing National Team]. Psychologie und Sport 4:123–134

20. Kellmann M, KW Kallus (1999) Mood, Recovery-Stress-State, and Regeneration. In: M Lehmann, Keizer H, Gastmann U, Steinacker JM (Eds) Overload, Fatigue, Performance Incompetence, and Regeneration. Plenum, New York, this volume

21. Kuipers H, Keizer HA (1988) Overtraining in elite athletes. Sports Med 6:79–92
22. Lehmann M, Baumgartl P, Wieseneck C, Seidel A, Baumann H, Fischer S, Spöri U, Gendrisch G, Kaminski R, Keul J (1992) Training - overtraining: influence of a defined increase in training volume vs. training intensity on performance, catecholamines and some metabolic parameters in experienced middle- and long-distance runners. Eur J Appl Physiol 64:169–177
23. Lehmann M, Gastmann U, Petersen KG, Bachl N, Seidel A, Khalaf AN, Fischer S, Keul J (1992) Training - overtraining: performance and hormone levels, after a defined increase in training volume vs. intensity in experienced middle- and long-distance runners. Br J Sports Med 26:233–242
24. Lehmann M, Knizia K, Gastmann U, Petersen KG, Khalaf AN, Bauer S, Kerp L, Keul J (1993) Influence of 6-week, 6 days per week, training on pituitary function in recreational athletes. Br J Sports Med 27:186–192
25. Lehmann M, Mann H, Gastmann U, Keul J, Vetter D, Steinacker JM, Häussinger D (1996) Unaccustomed high - mileage vs. intensity training—related changes in performance and serum amino acid levels. Int J Sports Med 17:187–192
26. Lehmann M, Lormes W, Opitz-Gress A, Steinacker JM, Netzer N, Foster C, Gastmann U (1997) Training and overtraining: an overview and experimental results in endurance sports. J Sports Med Phys Fitness 37:7–17
27. Liu Y, Mayr S, Opitz-Gress A, Zeller C, Lormes W, Lehmann M, Steinacker JM (1999) Muscular heat shock protein during training in highly trained rowers. J Appl Physiol, 86:101–104.
28. Lormes W, Lehmann M, JM Steinacker (1998) The problems to study plasma lactate. Int J Sports Med 19: 223.
29. Maassen N, Busse MW (1989) The relationship between lactic acid and work load: A measure for endurance capacity or an indicator of carbohydrate deficiency? Eur J Appl Physiol 58:728–737
30. Mickelson TC, Hagerman FC (1982) Anaerobic threshold measurements of elite oarsmen. Med Sci Sports Exerc 14:440–444
31. Morgan WP, Brown DR, Raglin JS, O'Conner PJ, Ellickson KA (1987) Psychological monitoring of overtraining and staleness. Br J Sports Med 21:107–114
32. Naessens G, Levevre J, Priessens M (1996). Practical and clinical relevance of urinary basal noradrenaline excretion in the followup of training processes in semiprofessional football players. Poster presented at: Conference on Overtraining in Sports. Memphis, TN.
33. Pedersen BK, Rohde T, Zacho M (1996) Immunity in athletes. J Sports Med Phys Fitness 36:236–245
34. Raglin JS, Morgan WP, Luchsinger AE (1990) Mood and self-motivation in successful and unsuccessful female rowers. Med Sci Sports Exerc 22: 849–853
35. Roth W, Hasart E, Wolf W, Pansold B (1983) Untersuchungen zur Dynamik der Energiebereitstellung während maximaler Mittelzeitausdauerbelastung. Med Sport 23:107–114
36. Secher NH (1993) Physiological and biomechanical aspects of rowing. Sports Med 15:24–42
37. Steinacker JM, Marx TR, Marx U, Lormes W (1986) Oxygen consumption and metabolic strain in rowing ergometer exercise. Europ J Appl Physiol 55:240–247
38. Steinacker JM (1993) Physiological aspects of training in rowing. Int J Sports Med 14:S3-S10
39. Steinacker JM, Laske R, Hetzel WD, Lormes W, Liu Y, Stauch M (1993) Metabolic and hormonal reactions during training in junior oarsmen. Int J Sports Med 14:S24-S28
40. Steinacker JM, Lormes W, Lehmann M, Altenburg D (1998) Training of Rowers before World Championships. Med Sci Sports Exerc 30:1158–1163
41. Steinacker JM, Lormes W, Kellmann M, Liu Y, Opitz-Gress A, Baller B, Günther K, Gastmann U, Petersen KG, Kallus KW, Lehmann M, Altenburg D. Training of Junior Rowers before World Championships (1998) Effects on performance, mood state and selected hormonal and metabolic responses. J Phys Fit Sports Med, in press.
42. Weicker H, Strobel G (1997) Endocrine regulation of metabolism during exercise. In: The physiology and pathophysiology of exercise tolerance. JM Steinacker, SA Ward (eds) Plenum London New York: 113–121
43. Wittert GA, Livesey JH, Espiner EA, Donald RA (1996) Adaptation of the hyothalamopituitary-adrenal axis to chronic exercise stress in humans. Med Sci Sports Exerc 28: 1015–1019
44. Urhausen A, Kullmer T, Kindermann W (1987) A 7-week follow-up study of the behavior of testosterone and cortisol during the competition period in rowers. Eur J Appl Physiol 56:528–533
45. Vervoorn C, Quist AM, Vermulst LJM, Erich WBM, deVries WR, Thijssen JHH (1991) The behavior of the plasma free testosterone/cortisol ratio during a season of elite rowing training. Int J Sports Med 12:257–263.

7

NEUROMUSCULAR AND MOTOR SYSTEM ALTERATIONS AFTER KNEE TRAUMA AND KNEE SURGERY

A New Paradigm

Jürgen Freiwald,[1] I. Reuter,[2] and Martin Engelhardt[1]

[1]Orthopedic University Hospital Friedrichsheim
J.W. Goethe University Frankfurt/Germany
[2]Mapother House, King's College Hospital, London, United Kingdom

1. INTRODUCTION

Knee trauma and operations have a high incidence and economic relevance (104). They are among the most frequent injuries especially in sports. Recently, both diagnostic possibilities (such as MRT) and the surgical spectrum (meniscus suture, cartilage cell breeding) have increased. Internationally, an increase in minimal-invasive procedures can be observed, supported by improved surgical equipment. Rehabilitation has changed, too, in parallel to the technical and surgical developments—it has been characterized by early-functional rehabilitation concepts since the early 90s. In spite of advances in diagnostics, surgical procedures and physiotherapeutic rehabilitation, there are short and/or long-term changes in motoric function following trauma and surgery. The subject matter of Motor System has nothing in common with the subject matter of movement. "In this way, there is a clear distinction between the totality of all regulating processes and functions on the one hand, and the adverse outcomes of these processes, human movement, on the other hand." (Marhold 1995 from Beyer 1992, 425). > Deficient activation of the musculature is seen, especially the knee extensors, consecutive atrophy—especially of the M. vastus medialis—and coordinative changes may occur. For posttraumatic and postoperative changes—for both the deficits and the therapy- or training-related adaptations—the explanations are meager. It is therefore urgently necessary to provide explanations in order to establish adequate treatment. The following questions must be answered:

1. How are the receptors in the knee joint supplied and what structures may be mechanically damaged in interior knee trauma or following surgical procedures?

Overload, Performance Incompetence, and Regeneration in Sport, edited by Lehmann *et al.*
Kluwer Academic / Plenum Publishers, New York, 1999.

2. What metabolic changes occur in conjunction with knee trauma or surgical procedures?

3. Are there any nervous functions disrupted in interior knee trauma or surgical procedures?

4. What changes can be measured in complex motor functions after knee injury and after surgical treatment?

5. Is there a pattern in the measurable changes?

6. What therapeutic consequences arise for rehabilitation?

2. HOW ARE THE RECEPTORS IN THE KNEE JOINT SUPPLIED AND WHAT STRUCTURES MAY BE MECHANICALLY DAMAGED IN INTERIOR KNEE TRAUMA OR FOLLOWING SURGICAL PROCEDURES?

Recently the purely biomechanical way of considering the structures of the knee joint has been altered in favour of a complex view taking functional relationships and neurophysiological knowledge into account.

The basis of this new vision was the intra- and extraarticular identification of receptor structures (Figure 1).

The most important receptors are the muscle spindles, Golgi tendon spindles and lamellar bodies, the Golgi, Ruffini and Pacini corpuscles found in the articular tissues, and the free nerve endings (8, 20). The capability of the receptors to be sensitive to specific in-

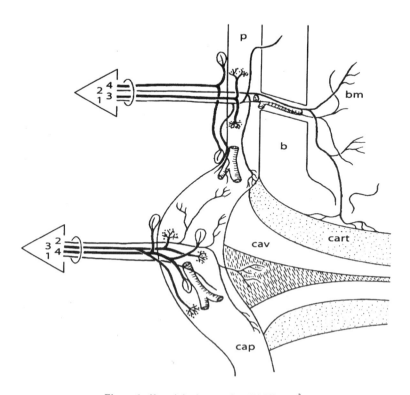

Figure 1. Knee joint innervation (Wolff 1996[2], 30).

formation is important for the physiological function of the knee joint (9, 19, 21, 32, 34, 49–51, 53, 131). It is agreed that the muscle and tendon spindles are indicators of changes in length and tension of the musculature and contribute essentially to kinaesthetic sensations. The importance of the joint receptors, on the other hand, remains controversial among scientists (4, 8, 11, 14–15, 17–18, 23, 25–26, 30–31, 33, 35–36, 40, 46–47, 49–59, 61–79, 81–84, 86–101, 103, 105–106, 109, 113–129, 130–133, 136–141). Due to the observation that damage to the joint receptors results in the majority of patients in considerable impairment of proprioreception and that the extensor musculature of the upper thigh is reduced following knee trauma to an extent which exceeds inactivity atrophy, the joint receptors have been assigned great importance (36, 42, 61, 137–138).

The most frequently-used division of joint receptor types today is based on Freeman and Wyke, who in 1967 described four different types of mechanoreceptors in the knee joint (34):

- Type 1: Ruffini bodies
- Type 2: Vater-Pacini bodies
- Type 3: Golgi organellae
- Type 4: Free nerve endings

Type 1 - Ruffini Bodies. The Ruffini bodies are mechanoreceptors with complex structure. These are, after the free nerve endings, the second most frequent receptor type in the knee joint. The axon is myelinized and has a conductance rate of 10–20 m/sec. Ruffini bodies are arranged in "clusters" of three to six bodies and are supplied by nerve fibers with a thickness of 5–10 µm, whereby broad variation may be seen (141). Ruffini bodies are slow-adapting receptors sensible for tension with a low stimulation threshold (95).

Type 2 - Vater-Pacini Receptors. This receptor type has been found in the fibrous knee joint capsule, Hoffa's infrapatellar fat body, the vascularized area of the medical meniscus, at the borderline between the ligament structures and the boney area and in the synovial sheath (4, 57, 51, 115). The conical Pacini receptors have a capsule consisting of several layers with a diameter of 400 µm to 1000 µm. The afferent axons of the Vater-Pacini organs are myelinized and have a diameter of about 6–10 µm; the axons of the lamellar bodies have a diameter of 3–6 µm. The conductance rate is between 25 and 50 m/s. The function of the Pacini bodies and the related nerve supply is the rapid uptake and transport of information from the afferent-supply tissues. They are inactive in immobile joints, likewise when the knee joint is moved at a constant velocity. They become sensor-active in the case of accelerations or decelerations. Vater-Pacini sensors have a low mechanical stimulation threshold and adapt rapidly; they act as "dynamic mechanoreceptors (139).

Type 3 - Golgi Receptors. The Golgi tendon organs in the muscle-tendon bond are identical to those in other tissues of the knee joint. Golgi organs are similar to the Ruffini bodies, and, as long as they are localized in the joints, no differentiation between them is made by some authors (95). Golgi organs are spindle-formed, measure about 600 × 100 µm in diameter and are enveloped in a capsule of connective tissue. The Golgi organs are supplied by one to six thick, myelinized nerve fibers (diameter 10–20 µm) with a conductance rate of about 75 m/s, which are designed Ib-fibers (66, 95, 112, 141). Golgi organs are found only in the area of the joint capsule, which is strengthened by ligaments or muscle tendons. Golgi organs are also found in the medial and lateral side ligament, the anterior and posterior cruciate ligament, the medial meniscus, in the lateral meniscus and in the muscle-tendon configuration of the patellar tendon (68, 141). The tendon organs are arranged in a series and measure the tension of the muscle tendon complex. The axon ends are in direct contact with collagen fiber bundles. They inhibit the muscular tension development in increasing muscle tension via activation of the Ib-afferents and only slowly adapt to mechanical stimuli. There is no agreement concerning the mechanical stress. Stress is used in this article to define external influence affecting the organism leading to internal demands. In patients with knee trauma, the capacity to cope with these demands may differ widely depending on the trauma, reconstruction and healing, required to elicit inhibiting afferents. The earlier conception that considered only a high mechanical (traction) stress as adequate stimulus to elicit Ib-afferents (compare 22, 141) is no longer tenable. More recent studies have shown that the afferents elicited by Golgi organs can be elicited by even minimum contraction strengths, such as that produced by the contraction of only one single muscle fiber (113). Whereas the muscle spindles measure primarily the length of the musculature, the Golgi tendon organs register the tension of the tendons or the muscle-tendon complex (9). The musculature therefore has two back-coupling systems available, a length-control system with muscle spindles and a tension-control system with the tendon organs as sensors. Golgi organs measure the ligamentary tension and in extreme biomechanical positions elicit increased Ib-afferents, which inhibit the α-motoneurons. This prevents further tension development and protects the structure involved. The overexertion protection of the muscle-tendon complex which can be deduced from this probably plays only a secondary role in the functional importance of the Golgi tendon organs. The Golgi tendon organs react to even low tensions which would not be any danger for the strength-developing muscles and may have not only inhibiting but also promoting influence (20). The reception threshold is variable and depends on both the cell milieu and on the multiconvergent-sensory integration. While the length-control system (muscle spindle system) is limited essentially to its own muscle and its antagonists, the muscle tone of the entire extremity is co-regulated by the tension-control system of the Ib afferents (9). The often postulated protective effect may function, if at all, only in slow-developing tendomuscular tension, since the nerve conductance rates are too slow or the latency times too long for effective protection of the muscle-tendon complexes (22, 86, 95, 115). Moreover, not all tissues are supplied with Golgi organs. The increasing inhibition of homonymic motoneurons with increasing muscular tension refers not only to the inhibition of tension development in the muscle-tendon complex in question. They may have promoting action on all flexors of the corresponding extremity and inhibiting action on the extensors (4). The importance of the Golgi system probably lies in control of the contraction strength for smooth control of the movement. This includes the task-appropriate pretensing of the musculature—which physiologically is of preventive importance. The knee joint or even the entire extremity is stabilized by appropriate pre-tensing of the musculature, which is created via central influences along with the Ib-afferent by afferent signals of the Group II λ-motoneurons (68).

Type 4 - Free Nerve Endings. The free nerve endings are receptors which are not encapsulated. The ends have multiple thick branches. The axons are thin and without myelinized sheath (about 1–2 μm in diameter). Other free nerve endings have a thin myelinized sheath with an axon diameter of about 2–4 μm (90–92). Free nerve endings occur frequently in "networks" with very thin, sensitive nerve endings of diameter about 0.5 to 1.5 μm (68, 141). The nociceptorily conducting fibers have a conduction velocity of 0.5 m/s to 30 m/s, whereby the fibers with conductance less than 2.5 m/s are frequently without myeline and are acribed to the Group IV fibers. The fibers which have a myeline sheath have a conductance velocity between 2.5 and 30 m/s and are ascribed to the Group III fibers (9, 92, 98). Free nerve endings are the most numerous receptors in the knee joint. They are frequently associated with vascular branches (95) and occur in the Stratum fibrosum of the capsule, the fat body of the plicae, the Villi synovials (8, 81), the inner and outer third of the meniscus, the perimeniscal tissue and the cruciate ligaments, especially near insertion (95, 54–58, 11–117). Nociceptive sensorics is one essential function of the free nerve endings. Free nerve endings usually have a high mechanical stimulation threshold. This is exceeded only in abnormal mechanical distortions and can influence motorics by eliciting reflexes (such as "giving way"). Beyond the capacity for mechanical information reception, the free nerve endings possess chemoreceptive properties which are of great clinical and functional importance, since each trauma and each chronic-degenerative joint disease is associated with inflammatory changes in (joint) metabolism (24). Similar to mechanical influences, the concentration of chemical signal substances must be very high to raise the receptor potential (ca. 100 mV) above the treshold value for action potential formation (ca.-70 mV) (95). The literature reports the existence of "silent receptors" which are first involved in the feedback process in conjunction with metabolites released in inflammatory processes. Schaible and Grubb showed in 1993 that inflammatory processes can greatly reduce the treshold values for excitability of the nerve endings (109). Similar results were reported by Kniffki et al. in 1979 and Mense in 1988 and 1995 (79, 90, 93). Due to the close coupling of the afferents (Group III and IV) of the free nerve endings with the α-efferents of the muscle spindles, the theoretical and working values of the musculature (among them pre-tensing, "muscular stiffness", elasticity of the musculature) and the adaptive capability of the musculature are co-determined by the mechanical and chemical reception of the free nerve endings. The extent to which the "adaptive pre-adjustment of the musculature" is altered by α-motoneuron activation, which is influenced by the perception of mechanical and metabolic stimuli of the free nerve endings among other factors, has not been clarified in all details. Based on current knowledge, however, such a coupling can be assumed (compare 70–71), especially since Hex et al. found in 1988 that 41% of the α-efferents, but only 14% of the α-efferents react to mechanical stimulation of the joint afferents (59). In 1986, Johansson et al. were able to achieve an even higher sensitivity of the α-efferents of 93 % after electrical stimulation of the posterior articular nerve (76).

2.1. Extraarticular Receptors, Muscle Spindles

The muscle spindle afferents are also important for the function of the knee joint. Each joint movement leads to changes in tension and length of the extraarticular musculature. The muscle spindles are only a few millimeters in size and are taken to be the receptors with the highest degree of complexity. They consist of receptive and activatable segments (88).

The muscle spindles are parallel and serial to the working musculature, frequently neighbouring tendenous muscle fascia. The muscle spindles, with 4–12 muscles within the muscle spindle, are surrounded by a capsule. Differentiation is made between nuclear fibers and nuclear bag fibers. They are differently enervated. The afferent nerve fibers of the muscle spindles, the so-called "anulospiral ending" is wound several times around the central of the intrafusal fiber. There is a Ia fiber in each muscle spindle. This contains myeline and has a diameter of 16–17 μm and conduction velocity of about 100 m/s (70 m/s-120 m/s) (compare 9, 11). Because they are supplied by the Ia fibers, the anulospiral endings are also called primary sensitive endings. Many, of not all muscle spindles have a second expansion-sensitive innervation. Their afferent fibers are thin (Group II fibers, diameter 8–9 μm). The receptor structures innervated by Group II fibers, which are connected to the spinal ganglia cells, are designated secondary muscle spindle endings (9). The Ia afferents stimulate homonym synapses (motoneurons). The Group II afferents also have a stimulating effect on homonym motoneurons, but their effect is multisegmental and similar to the transmission of those afferents which can elicit flexoreflexes (9, 106). Independent of the original muscle, they have promoting effect on all flexors of one extremity and inhibitory effect on all extensors. Thus, the muscle spindles affect the control of the entire extremity (87). The efferent innervation of the muscle spindles is effected by small α fibers. The corresponding efferent axons have a diameter of about 2–8 μm with a conduction velocity of about 50 m/s. At the resting length, the primary muscle spindles (I1 fibers) discharge. When the musculature is expanded, the Ia afferences increase. Functionally, this means that the muscle is stimulated to develop more tension in expansion of the muscle spindle receptors via Ia fiber afferences to offset a change in length. In addition to expansion of the extrafusal musculature and also the intrafusal musculature coupled to it, there is a second means of stimulating the muscle spindle endings. The intrafusal muscle spindles are capable of contracting themselves. The contraction command is elicited via the α-motoneurons. By contraction of the intrafusal muscle spindles, their central segment is expanded and leads likewise to afferent action potentials in the Ia fibers with a coupled increase in extrafusal muscle tension (112). The α-motoneuron system receives information from widely different tissue areas. Cutaneous afferences as well as stimulation of sensitive receptors in the intra- and periarticular tissue (Pacini, Ruffini, Golgi, free nerve endings) act in addition to central influences on preceding interneurons or the muscle spindle system (13, 45, 67). The contraction command via the α-motoneurons and the attendant change in the (pre)tensing of the extrafusal musculature can be elicited both by the afferences from the musculature itself and by afferent stimuli from other tissues in the knee joint and the periarticular tissue (67). Added to this are central control strategies and influences, such as stress and worrying circumstances in life (121). Any damage, any affection which is unknown to the Patient (such as prearthrotic joint processes) and any surgical procedure changes the sensory inflow and thus the α-efferences. The changes affect both static-coordinated (postural) and dynamic-coordinated, muscular activities (13, 48, 53, 66, 68–69, 141).

The muscle spindles act as pre-tensing regulators of the musculature and influence the motoneurons of the agonistic and antagonistic musculature either directly or via interneurons (66–76, 123, 126–127). Muscle expansion and the velocity of muscle expansion are adequate stimuli for Ia fibers (primary muscle spindle endings). The Ia fibers correspondingly conduct information on dynamic and static changes of the musculature.

The secondary afferent fibers, on the other hand, are less sensitive to dynamic changes, they react primarily to static stimuli. Corresponding to their innervation, the nuclear bag fibers function as differential sensors to indicate dynamic changes, while the nuclear chain fibers react as proportional sensors to static changes (105–106).

2.2. Traumatic and Intraoperative Receptor Damage

Receptors located both in the skin and in deeper layers may be damaged by knee trauma and surgical incision. With minimal-invasive procedures (arthroscopy), however, no serious damage to receptor structures is possible—sensitive structures may be stretched from being held with hooks. The influences of the blood blockage on sensitive organella and their afferent conduction pathways have not been clarified. It is known that the musculature reacts to blood blockage of more than 1-hour duration with massive functional loss and partial destruction (2, 63) and the threshold of nociception is reduced (67, 92).

3. WHAT METABOLIC CHANGES OCCUR IN CONJUNCTION WITH KNEE TRAUMA OR SURGICAL PROCEDURES?

Any trauma and any surgical procedure leads to an inflammatory reaction. The inflammation metabolites are received by chemosensitive free nerve endings (Type IV). The inflammation reaction may become chronic if the knee trauma does not heal completely or in the case of irreparable cartilage damage. There are typical spinal adaptations with demands on the flexor motoneurons and inhibition of the extensor motoneurons (79, 83). This evolution-related, phylogenetically valuable reaction leads to the clinically recordable and, using biomechanical methods, measurable weakening of the knee-extensor musculature and contractures of the knee-flexing musculature (see Figures 6, 9a and 9b).

4. WHAT NERVOUS FUNCTIONS ARE DISRUPTED IN INTERIOR KNEE TRAUMA OR SURGICAL PROCEDURES?

In order to answer question 3, we will start a "round trip" to check the intactness of the peripheral and central nervous system responsible for lower limb movement. The nervous system includes the peripheral and central nervous system. We assessed the peripheral

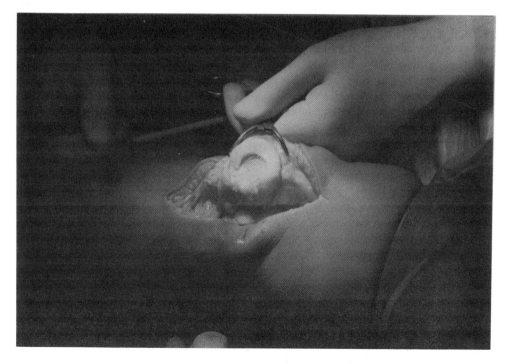

Figure 2. Situs by cartilage damage.

afferent pathways, the peripheral efferent pathways, the proximal segments of the peripheral nerves and the central processing.

4.1. Afferent Pathway

We performed sensory nerve conduction studies using surface electrods and examined the sural and peroneal nerve. There were no statistically significant differences between the injured and noninjured leg neither pre- nor postoperatively.

4.2. For Studying of the Proximal Nerve Segment We Conducted F-Wave and H-Reflex Measurements

The peroneal, tibial and femoral nerve were examined. The recordings did not reveal any differences between injured and noninjured leg neither pre- nor postoperatively.

4.3. Efferent Pathway

We performed motor nerve conduction studies to evaluate the function of the peripheral efferent pathway of the tibial, peroneal and femoral nerve. The recordings did not differ statistically significant between the injured and noninjured limb neither pre- nor postoperatively.

4.4. Central Pathway

We studied the central motor conduction by using transcranial magnetic stimulation and recorded the elicited compound muscle potential from the tibial muscle. The results were in normal range pre- and postoperatively and without statistically significant differences between injured and noninjured leg neither pre- nor postoperatively.

4.4.1. Narcosis Experiments. Studies by Bonebakker et al. (1996) and other authors show clearly that information can be received during procedures under surgically-required narcosis (10). Comparable processes are also plausible for proprioceptive information, which might influence postoperative rehabilitation in the sense of altered central-nervous control. In the future, we will have to pay more attention to the unconscious perception of proprio- and exterioreceptive stimuli under narcosis.

4.4.2. Neuromuscular Transmission. An altered EMG-strength ratio was conspicuous in all of the measurements which we performed in combination EMG and strength values (torques). Following knee trauma, a relatively higher central nervous activation of the knee extensor musculature is always necessary to create mechanical output (strength, torque) (35, 38). No unequivocal explanation for this phenomenon has yet been found (29) (Figures 3a and 3b).

In summary, it can be stated that there are no deficits in the afferent peripheral nervous pathway, central nervous processing and efferent nerve conduction pathways. There is no structural alteration of the nerval function. The measurable changes in muscle strength and the muscle atrophy seem to be due to an altered muscle activation. Hence this is not caused by structural alterations it represents an functional adaptation of the nervous system with changed activatability.

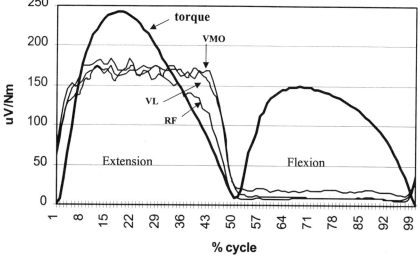

Figure 3a. EMG/torque relation. Isokinetic measurement 60°/s. VMO = M. vastus medialis oblique, VL = M. vastus lateralis, RF = M. rectus femoris. Rectified EMG; averaged curves from 22 patients with ACL rupture, not injured side.

5. WHAT CHANGES CAN BE MEASURED IN COMPLEX MOTORIC FUNCTIONS AFTER KNEE INJURY AND AFTER SURGICAL TREATMENT?

We measured a reduction in absolute strength application and active mobility after knee trauma and surgical treatment. There is a change in muscle selection and delays in neuromuscular reaction times.

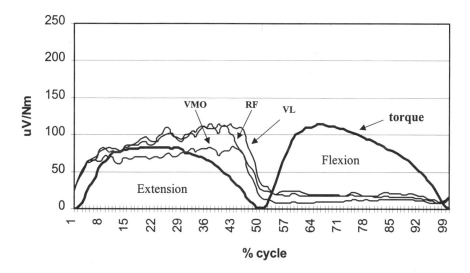

Figure 3b. EMG/torque relation. Isokinetic measurement 60°/s. VMO = M. vastus medialis oblique, VL = M. vastus lateralis, RF = M. rectus femoris. Rectified EMG; averaged curves from 22 patients with ACL rupture, injured and operated side. Week activation from M. vastus medialis (p=ns). Changed EMG/torque relation.

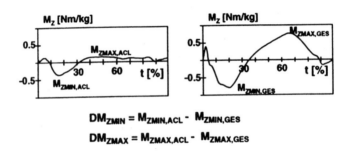

$$DM_{ZMIN} = M_{ZMIN,ACL} - M_{ZMIN,GES}$$

$$DM_{ZMAX} = M_{ZMAX,ACL} - M_{ZMAX,GES}$$

Figure 4. Moments in sagittal plane M_z (knee) from a ACL Patient 8 weeks p.o. (left side) and not injured leg (right side). Typical reduction on the operated side (Schmalz et al. 1998, 4).

5.1. Reduction of Absolute Strength Application (Torque)

After knee trauma and surgical treatment, strength (torque) is reduced. Only part of the strength reduction can be attributed to structural atrophy. In large part, reduced strength and torque values are due to central-nervous under-activation of the functional musculature (36) (Figure 4).

The reduction in strength (torque) arising in the central nervous system is not only measurable in the open system in isokinetic measurements, but also in the closed chain in gait analyses (1, 110) (compare Figure 5).

The changed absolute strength application leads to longer floor contact times in jumping exercises to be performed with as short floor contact as possible, (Tables 1a and 1b). Patients attempt to avoid especially the initial rapid-strength activation of the musculature; the cause here also appears to be an anticipative change in control of the musculature by the CNS.

5.2. Reduction of Active Mobility

While clinically measured, passive mobility can usually be completely restored during the first weeks after knee trauma and surgical treatment, the active mobility during everyday movement, such as gait, is limited. This remains significantly reduced even after a year as a result of altered central-nervous control for example after anterior cruciate ligament replacement (Figures 6a and 6b).

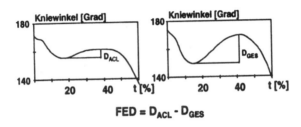

$$FED = D_{ACL} - D_{GES}$$

Figure 5a. Range of movement in sagittal plane (knee, stance phase) from a ACL patient 8 weeks after operation (left side) and from the noninjured knee (right side). Recognizable is a typical reduction on the operated side (Schmalz et al. 1998). FED=flexion-extension deficit.

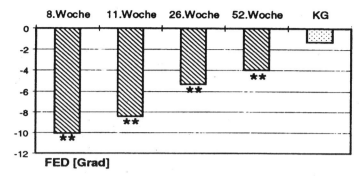

Figure 5b. Flexion-extension deficits. Patients group vs. controls (KG). Negative values correspond to deficits on the injured side (**: p = <0.01) (Schmalz et al. 1998, 5).

5.3. Changes in Muscle Selection

Selective atrophy of muscles is reported after knee trauma. The M. vastus medialis, which supposedly reacts especially sensitively to trauma of the knee joint, is cited particularly often. In the open chain, we could measure relative under-activations of the M. vastus medialis in isometric and isokinetic measurements in combination with EMG measurements in dependence on the type of trauma, but these were not significant (Figure

Table 1a. ACL, 12. week after bone-tendon-bone patellar plastic (n=10)[a]

	Control	Patients
t_{mit}	122.27	136.29
s	22.71	13.83
%	100	111.47
		+11.47
p		ns

[a]Control (n=10). Ground contact times [ms] - take off with boot feet. Target: minimal ground contact times [ms] (Freiwald 1996).

Table 1b. ACL, 12. week after bone-tendon-bone patellar plastic (n=10)[a]

	Control d	Control nd	Patients i	Patients u
t_{mit}	154.22	158.00	191.39	154.75
sd	24.80	20.19	52.80	16.69
%	100	102.45	80.86	100
%		+2.45	−19.14	
p-side	ns		<0.001	

[a]Control (n=10). Ground contact times [ms] - single leg jump. Target: minimal ground contact times [ms]. d = dominant, nd = not dominant, i = injured leg, u = uninjured leg (Freiwald 1996).

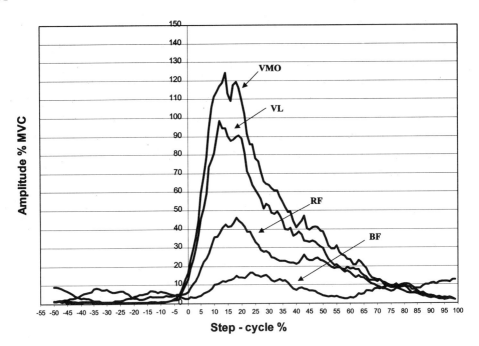

Figure 6a. Knee cartilage damage. Not injured leg, 63, 67 weeks postoperatively. Steps on a case (17.5 cm). 22 patients, 7 steps averaged. Normalized neuromuscular input from M. vastus medialis (VMO), M. vastus lateralis (VL), M. rectus femoris (RF) and M. biceps femoris (BF) (Freiwald et al., 1998).

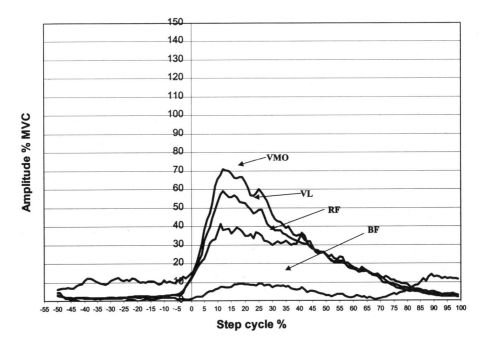

Figure 6b. Knee cartilage damage. Not injured leg, 63, 67 weeks postoperatively. Steps on a case (17.5 cm). 22 patients, 7 steps averaged. Normalized neuromuscular input from M. vastus medialis (VMO), M. vastus lateralis (VL), M. rectus femoris (RF) and M. biceps femoris (BF). Changed muscle selection, week activation from M. vastus medilias and M. vastus lateralis (Freiwald et al. 1998).

3a and 3b). In meniscus and cartilage injuries, we could not identify any selective under-activation of the M. vastus medialis (35, 38).

During step and jump exercises, we observed a clearly altered muscle selection, which was significant in dependence on the type of injury. There was a decrease in activation of the single-joint knee extensor M. vastus medialis and lateralis compared to the double-joint M. rectus femoris in all types on injuries (Fig. 7a and 7b), although the proportion of single-joint muscles was not greatly reduced compared to the other proportion.

5.4. Prolongation of Latency Times

The latency times between delivery of an optical signal and neuromuscular activation were prolonged after meniscal trauma (38). The latency times increased when the Patient was instructed to jump "in the injury mechanism". There was a dependency on the subjective expectations recorded in the questionnaires (Items; pain, lack of confidence fear) and changes in latency times. In addition to the latency times, the number of refused jumps also increased with an increase in negative expectations. The data are presently being evaluated. The signs of primary anticipatory processes controlled by the central nervous system are interesting in this connection.

6. IS THERE A PATTERN IN THE MEASURABLE CHANGES?

Surgical procedures are not part of the evolutionary reaction patterns of the organism. Procedures, which lead for example to capsule-ligament stabilization, are experienced by the Patient primarily as damage (27, 35). After knee trauma, the Patient changes his movement coordination. "Movement program" does not designate an analogy to kypernetic models. The activation patterns of the musculature are, in our understanding, part of a self-organizing system, are adapted to the reduced work capacity of receptor-supplied structures which are not totally functional. Our studies show that the Patient adapts biomechanical stress to the functional capacity of the knee joint mediated by the receptors and "estimated" in the central nervous system by variation of the adjustable program parameters mobility, absolute strength application, muscle selection and latency time. The estimation of functional capacity is formed by individual experience and evolutionary experience reflected in the specifics of neuronal structures and functional relays (5, 60). The measurable changes are meaningful for the organism. The altered neuromuscular activation is clear in fresh trauma or in the acute phase following surgery; the adaptations—depending on the severity and healing of the trauma—may, however, persist in weakened form.

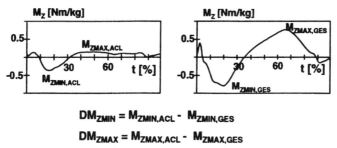

$$DM_{ZMIN} = M_{ZMIN,ACL} - M_{ZMIN,GES}$$

$$DM_{ZMAX} = M_{ZMAX,ACL} - M_{ZMAX,GES}$$

Figure 7. Range of movement in sagittal plane (knee, stance phase) from a patient 8 weeks after operation for the operated knee (left) and noninjured knee (right). Recognizable is the typical reduction of movement for the operated knee (Schmalz et al. 1998).

6.1. Sense of Reduction of Absolute Muscle Strength Application

In the closed chain, the Patient reduces the neuromuscular activation of the traumatized extremity and thus the floor reaction strengths affecting the traumatized joint (compare Figs. 5, 7a and 7b). By concurrent weight shifting during the floor contact phase up to the traumatized phase, the load is optimized and torques reduced. As in the closed chain, the neuromuscular activation is reduced in the open chain, too, leading to a reduction of torque. The changes in neuromuscular activation depend on the expected (anticipation) and really occurring (re) afferences and can differ clearly depending on the stress (e.g. open versus closed chain, joint velocity, etc.).

6.2. Sense of Limiting Active Mobility

The Patient spontaneously limits the active mobility after trauma. The limitations of mobility may persist as central-nervous controlled limitations during functional demands (compare Figs. 6a and 6b) and lead to consecutive-structural adaptation. The limitation of mobility improves the stability and control of the joint and reduces biomechanical stress, which is particularly pronounced in the area of end-grade movement.

6.3. Sense of Changes in Muscle Selection

We can measure changes in muscle selection; they appear subordinate to the reduction of neuromuscular activation capacity. The joint is additionally relieved by a change in muscle selection (e.g. M. vastus medialis weakness) only when the modifications by general neuromuscular under-activation capacity are not sufficient to avoid, for example, nociceptive afferences.

6.4. Sense of Changes in the Latency Time

Latency times are necessary to evoke behaviour following the eliciting signal. The biomechanical and biokinetic characteristics of the movement to be performed determine the temporal relationships (60). In our experiments, we used optically triggered selection reactions. It is known from the literature that if increasingly complex movements are demanded, the time from when the signal is given to performance of the movement increases (107). Due to the trauma, the sensory experiences which are coupled with a motor action are changed. Since motoric actions are not primarily motoric, but have a sensory-anticipative prestructure, the Patient must first learn to correctly predict the sensory consequences of defined motoric actions and make new assignments of their behavioral relevance (60). The latency times are prolonged as a consequence of this "new assignment" process.

When the adaptations of movement required in the sensory prediction—which are realized with changed latency and changes of neuromuscular activation—are not adequate, certain latent programs, such as jumping from a step or rapid running cannot be accessed (35, 38).

7. SUMMARY AND PROGNOSIS: WHAT THERAPEUTIC CONSEQUENCES ARISE FOR REHABILITATION?

The studies have shown that the prerequisites to correct proprioception are given after knee trauma and surgical treatment. The peripheral-afferent pathway, spinal and central stimulus processing and the peripheral-efferent pathway are intact after meniscal trauma and surgery. Altered peripheral-sensory perceptions (afferences) are caused by knee trauma or by the surgical treatment. Proprioception is not impaired, but rather changed by the trauma. After knee traumas and surgical treatment, adaptations are realized by a changed central-nervous control of the neuromuscular system. The Patient "creates his sensory world". He decides under anticipative prediction whether to perform movements (behaviour) or not. If the sensory consequences associated with the movement are considered intolerable, the behaviour is blocked (e.g. a jump from too-great a height). If the movement (behaviour) is estimated in the sensory anticipation to be tolerable, the neuromuscular system is appropriately activated. The biomechanical stress may be regulated via central-nervous adapted parametrization of the neuromuscular control to the actual functional capacity of the receptor-supplied tissues. For this, the absolute strength application, active mobility, muscle selection and latency times are adapted. The view presented here is a completely new paradigm. The deviations from the norm which have been stigmatized as "deficits" in biomechanical and neuromuscular parameters must be considered primarily as adaptations of the organism and interpreted as such. The Patient can create his own "sensory world". He does this within phylogenetically proven reaction mechanisms and within his biological possibility to adapt. The new paradigm can explain more than earlier views. The posttraumatic changes are supposed to protect traumatized structures. If knee traumas are treated surgically, the intraoperative traumatization is perceived as further trauma and the typical posttraumatic reactions follow. In spite of a biomechanically stabilized joint (e.g. anterior cruciate ligament treatment), there is a contradiction between sensory information (= additional, pronounced meniscal trauma) and biomechanical functional capacity. It becomes clear that operations are not included as part of phylogenetic development. In therapy, this paradigm switch must give pause for reflection. The patient has demonstrably no proprioceptic deficit, but rather proprioceptive information which has been adapted to the trauma and the surgical treatment. This altered information must be newly arranged in the access to basal central programs. Central programs must be accessed after surgical procedures and adapted to the changed peripheral information. The goal of therapy must be to re-access the latent movement capacities like walking, running and jumping, taking the current functional capacity into account. The patient, does not have to relearn these programs! In order to realize this, the necessary perquisites must be created. Since the Patient modifies the movement programs by this "sensory expectations", the essential afferent influences like pain, swelling and effusion must be prevented or relieved. Postoperative exercises must be so selected that the patient can collect positive sensory experiences (memory contents, somatosensory cortex) in close to everyday activities. Targeted use of mental forms of training (ideomotoric training forms) such as have been effective in sports and intraoperative positive suggestion are plausible. Physicians must give some thought to the intra- and perioperative prevention of negative conditioning sensory information. Procedures of peripheral, intraarticular anaesthetic and medication therapy of swelling, effusion and pain are plausible.

In the future, changes in neuromuscular and biomechanical parameters must be viewed under the paradigm of compensation, as adapted reactions e.g. after meniscal

trauma and deficient, non-adapted reaction e.g. after surgical procedures, and propriocep-
tive-induced, neuromuscular and biomechanical changes re-assessed.

REFERENCES

1. Andriacchi TP, Birac D (1993) Functional testing in the anterior cruciate ligament deficient knee. Clin Orthop Rel Res 288: 40–47
2. Appell HJ, Glöser S, Duarte JAR, Zellner A, Soares JMC (1993) Skeletal muscle damage during tourni-
   quet-induced ischemia. The initial step towards artrophy after orthopaedic surgery? Eur J Appl Physio 67:
   342–347
3. Appell HJ, Verdonck A, Duesberg, F, Windeck P (1991) Fehlende Ermüdung der Muskulatur bei Patienten
   nach Immobilisation - ein Paradoxon? Sportverletzung - Sportschaden (5): 205–207
4. Assimakopulos AP, Katonis PG, Agapitos MV, Exarchou EI (1992) The innervation of the human meniscus. Clinical Orthopaedics and Related Research 275, 2: 232–236
5. Balgo R (1998) Bewegung und Wahrnehmung als System. Hofmann Schorndorf
6. Bernstein NA (1987)2 Bewegungsphysiologie. Barth Leipzig
7. Beyer E (ed) (1992)2 Dictionary of Sport Science. Hofmann Schorndorf
8. Biedert RM, Stauffer E, Friederich NF (1992) Occurrence of free nerve endings in the soft tissue of the
   knee joint. The American Journal of Sports Medicine 20, 4: 430–433
9. Birbaumer N, Schmidt RF (1989) Biologische Psychologie, Berlin
10. Bonebakker AE, Bonke B, Klein MD, Wolters G, Stijnen T. Passchier J, Merikle PM (1996) Information
    processing during general anaesthesia: Evidence for unconscious memory. Memory & Cognition 24: 766–776
11. Boyd IA, Roberts TDM (1953) Proprioreceptive discharges from stretchreceptors in the knee joint of the
    cat. J. Physiology
12. Brand RA (1989) A neurosensory hypothesis of ligament function. Med Hypotheses (29) 4: 245–250
13. Brügger A (1980) Die Erkrankungen des Bewegungsapparates und seines Nervensystems. Stuttgart-New
    York
14. Burgness PR, Clark FJ (1969) Characteristics of knee joint receptors in the cat. J Physiol 203-B: 317–335
15. Burke D, Gandevia SC, Macefield G (1988) Responses to passive movement of receptors in joint, skin and
    muscle of the human hand. J Physiol [Br] (402): 347–361
16. Claus D (1993) Transkranielle Stimulation. In: Görg J and Hielscher H (eds) Evozierte Potentiale in Klinik
    und Praxis. Springer: 347–361
17. Corrigan JP, Cashman WF, Brady MP (1992) Proprioception in the cruciate deficient knee. J Bone Joint
    Surg 74-B: 247–250
18. DeAndrade JR, Grant C, Dixon ASJ (1965) Joint distension and reflex muscle inhibition in the knee. J
    Bone Joint Surg 47A: 313–322
19. Dietz V, Horstmann GA, Trippel M, Gollhofer A (1989) Human postural reflexes and gravity - an underwa-
    ter simulation. Neurscience Letters 106:350–355
20. Dietz V (1997) Neuronal Kontrolle automatischer funktioneller Bewegungsabläufe: Wechselbeziehung
    zwischen zentraler Programmierung und afferenter Information. In: Zichner L, Engelhardt M, Freiwald J
    (eds): Muskuläre Dysbalancen. Novartis Wehr: 59–69
22. Eccles JC (1969) The Inhibitory Pathways of the Central Nervous System. The Sherrington Lectures IX
    Springfield/III
23. Eckhardt R, Schaft HP, Puhl W (1994) Die Bedeutung der neuromuskulären Koordination für die
    sportliche Belastbarkeit des Kniegelenkes nach vorderen Kreuzbandverletzungen. Sportverletzung-Sport-
    schaden 8: 16–24
24. Edelson R, Burkes BT and Bloebaum RD (1995) Short term effects of knee washout for osteoarthritis. Am
    J Sports Med 23, 3: 345–349
25. Edin BB, Abbs JH (1991) Finger Movement Responses of Cutaneous Mechanoreceptors in the Dorsal
    Skin of the Human Hand. Journal of Neurophysiology 65, 3: 657–660
26. Elmqvist LG, Lorentzon R, Johansson C, Fugl-Meyer AR (1988) Does a torn anterior cruciate ligament
    lead to change in the central nervous drive of the knee extensors? Eur J Appl Physiol 58: 203–207
27. Engelhardt M (1997) Neuromuskuläre Veränderungen nach Kniegelenkstraumen und Operationen am
    Kniegelenk. Habitilationsschrift, Frankfurt
28. Engelhardt M, Reuter I, Freiwald J (submitted 1998) Is muscle atrophy after knee injury caused by reduced
    neural activation? Medicine, Sports and Science

29. Engelhardt M, Freiwald J (1997) EMG-kontrollierte Muskelrehabilitierung - Knieverletzungen. Sportverletzung-Sportschaden 11, 3: 87–99

30. Ferrell WR, Rosenberg JR, Baxendale RH, Halliday D, Wood L (1990) Fourier analysis of the relation between the discharge of quadriceps motor units and periodic mechanical stimulation of cat knee joint receptors. Experimental Physiology 75: 739–750

31. Ferrell, WR, Danvevia SC, McCloskey DI (1987) The role of joint receptors in human kinaesthesia when intramuscular receptors cannot contribute. J Physiol (386) 5: 63–71

32. Ferrell WR, Baxendale RH, Carnachan C, Hart IK (1985) The influence of joint afferent discharge on locomotion, proprioreception and activity in conscious cats. Brain Res (347): 41–48

33. Freeman MAR (1965) Treatment of ruptures of the lateral ligament of the ankle. J Bone Joint Surg 47-B: 661–668

34. Freeman MA, Wyke B (1967) The Innervation of the Knee Joint: An Anatomical and Histological Study in the Cat. J Anat 101, 3: 505–512

35. Freiwald J (1996) Neuromuskuläre Veränderungen des M. quadrizeps femoris nach akuten und chronischen Kniegelenksschädigungen. Habilitationsschrift Dortmund

36. Freiwald J (1992) Veränderungen von Umfangsmaßen, isometrischen und isokinetischen Kraftwerten nach Schädigungen des Kniegelenkes unter besonderer Berücksichtigung neurophysiologischer Ursachen. Dissertationsschrift, Dortmund.

37. Freiwald J, Engelhardt M, Gnewuch A (1998) Trainingstherapie auf der Basis der Motorikforschung und der philosophischen Erkenntnistheorie am Beispiel von Kniepatienten. In: Binkowski H, Hoster M, Nepper HU (eds) Medizinische Trainingstherapie in der ambulanten orthopädischen und traumatologischen Rehabilitation. Sport Consult Waldenburg: 9–19

38. Freiwald J, Engelhardt M, Huth D (1998) Veränderungen der neuronalen Ansteuerungsmuster der Beinmuskulatur nach Kniebinnentraumen. Poster presented at the 13th German-Austrian-Swiss Congress for Sportsorthopedics and Sport Traumatology, Munich

39. Freiwald J, Engelhardt M, Reuter I, Konrad P, Gnewuch A (1997) Die nervöse Versorgung der Kniegelenke. Wiener Medizinische Wochenzeitschrift. Themenheft "Kniegelenk" 23/24: 531–541

40. Freiwald J, Engelhardt M, Reuter I (1995) Der Einfluß von intraartikulär applizierten lokalen g und Training. Sankt Augustin: 245–250

41. Freiwald J, Engelhardt M (1994) EMG gestützte Funktionsanalysen nach vordren Kreuzbandplastiken. In: Schmidtbleicher D, Müller AF (eds) Leistungsdiagnostische und präventive Aspekte der Biomechanik. Sankt Augustin: 123–136

42. Freiwald J, Engelhardt M (1994) EMG-Einsatz in der Knierehabilitation. Praktische Konsequenzen. Rehabilitace a Fyzikalni Lekarstivi 374: 136–139

43. Freiwald J, Starischka S, Engelhardt M (1993) Rehabilitatives Krafttraining. Überlegungen zum Krafttraining - Neue Ansätze zur Anwendung und Diagnostik im klinischen Bereich. Deutsche Zeitschrift für Sportmedizin 44, 9: 368–378

44. Frisch H (1995) Programmierte Therapie am Bewegungsapparat. Springer, Berlin

45. Frisch H (1989) Programmierte Untersuchung des Bewegungsapparates. Springer-Verlag Berlin-Heidelberg-New York

46. Gardner E (1944) The distribution and termination of nerves in the knee joint of the cat. J Compu Neurol 80: 11–32

47. Goertzen M, Gruber J, Dellmann A (1992) Neurohistological findings after experimental anterior cruciate ligament allograft transplantation. Arch Orthop Trauma Surg (111) 2: 126–129

48. Gollhofer A, Scheuffelen C, Lohrer H (1993) Neuromuskuläre Stabilisation im oberen Sprunggelenk nach Immobilisation. Sportverletzung-Sportschaden (Sonderheft 1), 7: 23–28

49. Grigg P, Schaible HG, Schmidt RF (1986) Mechanical sensitivity of group III and IV afferents from posterior articular nerve in normal and inflamed cat knee. Journal of Neurophysiol 55, 4:635–643

50. Grigg P, Hoffman AH (1984) Ruffini mechanoreceptors in isolated joint capsule: response correlated with strain energy density. Somatosens Res 2: 149–162

51. Grigg P, Hoffmann AH (1982) Properties of ruffini afferents revealed by stress analysis of isolated sections of cat knee capsule. J of Neurophysiology 47, 1: 41–45

52. Grillner S, Hongo T, Lund S (1969) Descending monosynaptic and reflex control of Gamma-mononeurons. Acta Physiol Scand 75:592

53. Grüber J, Wolter D, Lierse W (1986) Der vordere Kreuzbandreflex (LCA-Reflex). Unfallchirurg 89: 551–554

54. Halata Z, Haus J (1989) The ultrastructure of sensory nerve endings in human anterior cruciate ligament. Anat Embryol 179:415–421

55. Halata Z (1988) Ruffini corpuscle - a stretch receptor in the connective tissue of the skin and locomotion apparatus. In: Hamann W, Iggo A (eds) Progress in Brain Research Vol 74: 221–229

56. Halata Z, Groth HP (1976) Innervation of the synovial membrane of the cats joint capsule. Cell Tissue Res 1969: 415–418

57. Haus J, Halata Z, Refior HJ (1992) Propriozeption im vorderen Kreuzband des menschlichen Kniegelenkes - morphologische Grundlagen. Z Orthop 130: 484–494

58. Haus J, Halata Z (1990) Innervation of the anterior cruciate ligament. International Orthopaedics (SICOT) 14: 293–296

59. He X, Proski U, Schaible HG (1988) Acute inflammation of the knee joint in the cat alter responses of flexor motoneurons to leg movement. J Neurophysiol (59) 2: 326–340

60. Hoffmann J (1993) Vorhersage und Erkenntnis. Hogrefe, Göttingen

61. Hörster G, Kediziora O (1993) Kraftverlust und -regeneration der Kniestreckmuskulatur nach Operationen am Kniebandapparat. Akt Sporttraumatol 23: 244–254

62. Hultborn H (1972) Convergence on interneurons in the reciprocal Ia inhibitory pathway to motoneurones. Acta Physiol Scand (Supplement) 375, 85: 1–42

63. Jacobson MD, Pedowitz RA, Oyama BK, Tryon G, Gershuni DH (1993) Muscle Functional Deficits after Tourniquet Ischemia. The American Journal of Sports Medicine 22, 3: 372–377

64. Jerosch J, Castro WHM, Hofstetter I, Bischof M (1994) Propriozeptive Fähigkeiten bei Probanden mit stabilen und instabilen Sprunggelenken. Deutsche Zeitschrift für Sportmedizin 45, 10: 380–389

65. Jerosch J, Hofstetter I, Reer R, Assheuer J (1994) Strain-related long-term changes in the minisci in asymptomatic athletes. Knee Surg Sports Traumatol Arthroscopy 2: 8–13

66. Johansson H, Sjölander P, Sojka P (1991) Receptors in the knee joint ligaments and their role in the biomechanics of the joint. Critical Reviews in Biomedical Engineering 18, 5: 341–368

67. Johansson H, Sojka P (1991) Pathophysiological mechanisms involved in genesis and spread of muscular tension in occupational muscle pain and in chronic musculoskeletal pain syndromes: A hypothesis. Medical Hypotheses 35: 196–203

68. Johansson H (1991) Role of knee ligaments in proprioception and regulation of muscle stiffness. Journal of Electromyography and Kinesiology I, 3: 158–179

69. Johansson H, Sjölander P, Sojka P (1991) A sensory role for the cruciate ligaments. Clinical Orthopaedics and Related Research 268: 161–178

70. Johansson H, Sjölander P, Sojka P (1990) Activity in receptor afferents from the anterior cruciate ligament evokes reflex effects on fusimotor neurones. Neurscience Research 8: 54–59

71. Johansson H, Lorentzon R, Sjölander P, Sojka P (1990) The anterior cruciate igament. A sensor acting on the (-muscle spindle systems of muscles around the knee joint. Neuro Orthop (9): 1–23

72. Johansson H, Sjölander P, Sojka P, Wadell I (1989) Reflex actions on the Gamma-muscle spindle systems of muscles acting at the knee. Neuroorthopedics 8: 9

73. Johansson H, Sjölander P, Sojka P (1989) Effects of electrical and natural stimulation of skin afferents on the gamma-spindle system of the triceps surae muscle. Neurosci Res (6) 6: 537–555

74. Johansson H, Lorentzon R, Sjöström M, Fagerlund M, Fugl-Meyer AR (1987) Sprinter and marathon runners. Does isokinetic knee extensor performance reflect muscle size and structure. Acta Physio Scand 130: 663–669

75. Johansson H, Sjölander P, Sojka P (1987) Fusimotor reflexes to antagonistic muscles simultaneously assessed by multi-afferent recordings from muscle spindle afferents. Brain Res (435) 1–2: 337–342

76. Johansson H, Sjölander P, Sojka P (1986) Actions on gamma-motoneurones elicited by electrical stimulation of joint aferent fibres in the hind limb of the cat. J Physio 375: 137–152

77. Katonis PG, Assimakopoulos AP, Agapitos MV, Exarchou EI (1991) Mechanoreceptors in the posterior cruciate ligament. Acta Orthop Scand 72, 3: 276–278

78. Kennedy JC, Alexander IJ, Hayes KC (1982) Nerve Supply of the Human Knee and its Functional Importance. Am J Sports Med 10, 6: 329–335

79. Kniffki KD, Schomburg ED, Steffens H (1979) Synaptic responses of lumbar Alpha-motoneurones to chemical algesic stimulation of skeletal muscle in spinal cats. Brain Res 160: 549–552

80. Konrad P (1996) Analyse von Belastungs- und Beanspruchungsindikatoren im Kunstturntraining - unter besonderer Berücksichtigung neuromuskulärer Messverfahren. Sport & Buch Strauß, Cologne

81. Langford LA, Schaible HG, Schmidt RF (1984) Structure and function of fine joint afferents; Observations and speculations. In: Hamann W, Iggo A (eds) Sensory receptor mechanisms. World scientific Singapore

82. LaRue J, Bard C, Fleury M (1995) Is proprioception important for the timing of motor activities? Can J Physiol Pharmacol (73) 2: 255–261

83. Lass P, Kalund S, LeFevre S, Arendt-Nielsen L, Sinkjaer R, Simonsen 0 (1991) Muscle coordination following rupture of the anterior cruciate ligament. Acta Orthop Scand 62: 9–14

84. Lentell G, Baas B, Lopez D (1995) The contributions of proprioceptive deficits, muscle function, and anatomilaxity to functional instability of the ankle. J Orthop Sports Phy Ther (21) 4: 206–215

85. Lorentzon R, Johansson C, Sjöström M, Fagerlund M, Fugl-Meyer AR (1988) Fatigue during dynamic muscle contractions in male sprinters and marathon runners: Relationship between performance, electromyographic activity, muscle cross-sectional area and morphology. Acta Physiology Scand 132: 531–536

86. Lundberg A, Lamgren K, Schomburg ED (1978) Role of joint afferents in motor control exemplified by effects on reflex pathways from Ib afferents. J Physio (Lond) 184: 327–343

87. Lundberg A, Malmgren K, Schomburg ED (1977) Cutaneous facilitation of transmission in reflex pathways from Ib afferents to motoneurones. J Physio (Lond) 265: 763–780

88. Matthews PBC (1972) Mammalian muscle receptors and their central actions. Arnold London

89. McNair PJ, Marshall RN, Marguire K (1995) Knee joint effusion and proprioception. Arch Phys Med Rehabil (76) 6: 566–568

90. Mense S (1995) Lokaler und übertragener Muskelschmerz. Phys Rehab Kur Med 5: 147–152

91. Mense S (1993) Nociception from skeletal muscle in relation to clinical muscle pain. Pain 54: 241–289

92. Mense S (1991) Physiology of nociception in muscles. J Manual Medicine 6: 24–33

93. Mense S (1988) Verhalten von Nozizeptoren im normalen und im entzündeten Muskel. In: Sprintge R, Droh R (eds) Schmerz und Sport Heidelberg: 199–206

94. Newell KM, Corcos DM (eds) (1993) Variability and Motor Control. Human Kinetic Publishers, Champaign

95. Nürnberger F (1997) Lokalisation und Funktion von Rezeptoren im Gelenk-Muskel-Complex. In: Zichner L, Engelhardt M, Freiwald J (eds) Muskuläre Dysbalancen. Novartis, Wehr: 24–38

96. O'Connor BL, Visco DM, Brandt KD (1993) Sensory nerves only temporarily protect the unstable canine knee joint from osteoarthritis. Evidence that sensory nerves reprogram the central nervous system after cruciate ligament transection. Arthritis Rheum (36) 8: 1154–1163

97. O'Connor BL, Palmoski MJ, Brandt KD (1985) Neurogenic acceleration of degenerative joint lesions. J Bone Joint Surg [Am] (67) 4: 562–572

98. Paintal AS (1967) A comparison of the nerve impulses of mammalian non medullated nerve fibres with those of smallest diameter medullated fibres. J Phys (Lond) 193: 523–533

99. Pitman MI, Nainzadeh N, Menche D (1992) The intraoperative evaluation of the neurosensory function of the anterior cruciate ligament in humans using somatosensory evoked potentials. Arthroscopy (8) 4: 442–447

100. Pope CF, Cole KJ, Brand RA (1990) Physiologic loading of the anterior cruciate ligament does not activate quadriceps or hamstrings in the anesthesized cat. The American Journal of Sports Medicine 18, 6: 595–599

101. Pope MH, Johnson RJ, Brown DW, Tighe C (1979) The role of the musculature in injuries to medial collateral ligament. J Bone Joint Surg 61: 398–402

102. Popper K (1994) Logik der Forschung. Mohr Tübingen

103. Portr R, Lemon R (1995) Corticospinal Function an Voluntary Movement. Clarendon Press Oxford

104. Renström PAFH (ed) (1993) Sports Injuries. Blackwell Oxford.

105. Reuter I, Engelhardt M, Freiwald J (1994) Sensorische Rückmeldungen aus arthronalen Systemen als Steuerungsvoraussetzungen der Muskulatur. In: Zichner, L, Engelhardt M, Freiwald J (eds) (1994) Die Muskulatur. Sensibles, integratives und meßbares Organ. Ciba Geigy Wehr: 41–52

106. Reuter I, Engelhardt M, Freiwald J (1994) Steuerung der Muskulatur durch sensorische Rückmeldung. TW Sport und Medizin 6, 3: 181–184

107. Rosenbaum DA, Gordon AM, Stillings NA, Feinstein MH (1987) Stimulus-response compatability in the programming of speech. Memory & Cognition 15: 372–393

108. Rovere GD, Adair DM (1983) Anterior cruciate-deficient knees: a review of the literature. Am J Sports Med (11): 412

109. Schaible HG, Grubb BD (1993) Afferent and spinal mechanisms of joint pain. Pain 55: 5–54

110. Schmalz T, Blumentritt S, Wagner R, Gokeler A (1998) Ganganalytische Verlaufsuntersuchung patellasehnenversorgter Rupturen des vorhandenen Kreuzbandes. Phys Rehab Kur Med 9: 1–8

111. Schmidt RA (1988) Motor control and learning: A behavioral emphasis. Human Kinetics Champaign

112. Schmidt RF (ed) (1987) Grundriß der Neurophysiologie. Springer Verlag Berlin

113. Schomburg ED (1997) Spinale Mechanismen zur Steuerung neuromuskuärer Balance. In: Zichner L, Engelhardt M, Freiwald J (eds) Neuromuskuläre Dysbalancen Novartis Wehr: 39–57

114. Schomburg ED (1991) The role of nociceptive afferents and enkephalins in spinal motor control. In: Wernig A (ed) Plastisticity of Motoneural Connection. Elsevier Amsterdam: 345–353

115. Schmburg ED (1988) Zur Funktion nozirezeptiver Afferenzen in der spinalen Motorik. In: Spintge R, Droh R (eds) Schmerz und Sport. Berlin-Heidelberg: 207–219

116. Schultz RA, Miller CD, Kerr C, Micheli L (1984) Mechanorezeptoren in human cruciate ligaments. J Bone Joint Surg 66-A: 1072–1076

117. Schutte MJ, Dabezies EK, Zimny ML (1987) Neural Anatomy of the Human Anterior Cruciate Ligament. J Bone Joint Surg [Am] (69): 243–247

118. Scott DT, Ferrell WR, Baxendale RH (1994) Excitation of soleus-gastrocnemius gamma-motoneurones by group II knee joint afferents in suppressed by group IV joint afferents in the decerebrate, spinalized cat. Exp Physiol 79: 357–364

119. Shakespeare DT, Stokes M, Sherman KP (1985) Reflex inhibition of the quadriceps after meniscectomy: lack of association with pain. Clin Physiol 5: 137–144

120. Shelbourne KD, Nitz P (1990) Accelerated rehabilitation after anterior cruciate ligament reconstruction. Am J Sports Med 18: 292–299

121. Simons DG (1988) Myofascial pain syndrome due to trigger points. Reh Medicine St. Louis: 686–723

122. Sinkjaer R, Arendt-Nielsen L (1991) Knee stability and muscle coordination in patients with anterior cruciate ligament injuries. An electromyographic approach. J Electromyography Kinesiol 3, 1: 209–217

123. Sjölander P, Djupsjöbacka M, Johansson H, Sojka P, Lorentzon R (1994) Can receptors in the collateral ligaments contribute to knee stability and proprioception via effects on the fusimotor-muscle-spindle system? Neuro-Orthopedics 15:65–80

124. Skinner HB, Barrack RL (1991) Joint Position Sense in the Normal and Pathologic Knee Joint. Journal of Electromyography and Kinesiology (1) 3: 180–190

125. Snyder Macker L, DeLuca PF, Williams PR (1994) Reflex inhibition of the quadriceps femoris muscle after injury or reconstruction of the anterior cruciate ligament. J Bone Joint Surg [Am] (76) 4: 555–560

126. Sojka P, Sjölander P, Johansson H, Dupsjöbacka M (1991) Influence from stretch-sensitive receptors in the collageral ligaments of the knee joint on the (-muscle spindle systems of flexor and extensor muscles. Neurosci Res 11: 55–62

127. Sojka P, Sjölander P, Johansson H, Djupsjöbacka M (1989) Fusimotor neurons can be reflexly influenced by activity in receptor afferents from the posterior cruciate ligament. Brain Res 483: 177

128. Spencer JD, Hayes KC, Alexander IJ (1984) Knee joint effusion and quadriceps reflex inhibition in man. Arch Phys. Med Rehabil (65) 4: 171–177

129. Stöhr M and Bluthardt M (1993) Atlas der klinischen Elektromyographie und der Neurographie. Kohlhammer, 3. Auflage

130. Stokes M, Young A (1984) The contribution of reflex inhibition to arthrogenous muscle weakness. Clinical Science 67: 7–14

131. Wißmeier T, Kutter T, Hülser PJ (1997) Der H-Reflex - eine neue Möglichkeit der Kontrolle von Funktionsparametern in der Behandlung von Bandverletzungen. Beispiel: Vorderes Kreuzband. In: Zichner L, Engelhardt M, Freiwald J (eds). Muskuläre Dysbalancen. Novartis Wehr: 133–164

132. Wojtys EM, Juston LJ (1994) Neuromuscular performance in normal and anteiror cruciate ligament-deficient lower extremities. Am J Sports Med 22, 1: 89–104

133. Wolff HD (1996) Neurophysiologische Aspekte des Bewegungssystems. Springer Berlin

134. Wollny R (1993) Stabilität und Variabilität im motorischen Verhalten. Meyer & Meyer Aachen

135. Wulf G (1994) Zur Optimierung motorischer Lernprozesse. Hofmann Schorndorf

136. Wyke B (1967) The Neurology of Joints. Ann R Coll Surg Engl 41:25–50

137. Young A, Stokes M, Iles JF (1987) Effects of joint pathology on muscle. Clinical Orthopaedics and Related Research 219, 6: 21–27

138. Young A, Stokes M (1986) Reflex inhibition of muscle activity and the morphological consequences of inactivity. In: Saltin B (ed) International Series of Sport Sciences, Vol. 16, Biochemistry of Exercise VI. Human Kinetics, Champaign: 531–544

139. Zimny ML (1988) Mechanoreceptors in articular tissues. The American Journal of Anatomy 182: 16–32

140. Zimny ML, Schutte M, Dabezies E (1986) Mechanoreceptors in the human anterior cruciate ligament. Anar Rec (214) 2: 204–209

141. Zimny ML, Wink CS (1991) Neuroreceptors in the tissues of the knee joint. Journal of Electromyography and Kinesiology I, 3: 148–157

8

MOOD, RECOVERY-STRESS STATE, AND REGENERATION

Michael Kellmann[1] and Konrad W. Kallus[2]

[1]Institute of Sport Science
Section for Sportpsychology
Potsdam University
Am Neuen Palais 10
14469 Potsdam, Germany
[2]Department of Psychology
Section for Applied Psychology
Karl-Franzens University Graz
Universitätsplatz 3
A-8010 Graz, Austria

1. INTRODUCTION

Stressful high intensity training periods are necessary to obtain high performance in sports. However, the simple rule 'the more - the better' does not apply in this context. A lot of studies clearly showed that systematic recovery periods in the training process are necessary to prevent an overtraining syndrome or staleness (39), and to obtain overreaching for further performance improvement (28). Sportmedical assessment of athletes' training states are used as valuable tools to determine the necessary amount of training and recovery in preparing an athlete for peak performance. In addition, authors using the Borg's Rating of Perceived Exhaustion (RPE, 3, 37) or the Profile of Mood States (POMS, 31) clearly showed that changes in training load are reflected in subjective states and in the mood of athletes (34, 36, 38, 40). In particular, studies dealing with the POMS have demonstrated that the typical 'Iceberg-Profile' (33, 35) of the well-trained athlete changes during a phase of intensive training and deteriorates during overtraining (39). Morgan at al. (34) reported mood changes in swimmers during the season. Early in the season, swimmers displayed the 'Iceberg-Profile', a profile indicative of positive mental health that is associated with successful athletic performance. During overtraining, however, mood disturbances significantly increased and were accompanied by a profile reflecting diminished mental health. After training stimulus was significantly reduced following a taper, the swimmers again exhibited

Overload, Performance Incompetence, and Regeneration in Sport, edited by Lehmann *et al.*
Kluwer Academic / Plenum Publishers, New York, 1999.

the original 'Iceberg-Profile'. More recently, the existence of a dose-response relationship was demonstrated between training volume and mood disturbances (39). Increases in training volume parallel corresponding elevations in mood disturbance (e.g., greater anger, depression, tension, fatigue, and less vigor and well-being). Mood improvements occur if the training volume is reduced (34, 36, 38, 40). Stress models have been proposed to account for the effects of overtraining considering intensive training as a kind of chronic intermittent stressor. Meyers and Whelan (30) used this approach to show that training stressors and non-training stressors possibly contribute to the athletes' state in a joint manner that allows to explain the large variability in the psychological effects of intensive training. Their approach is primarily based on the psychological stress model developed by Lazarus and co-workers (27). Our conception integrates effects of common private, and training stressors using a biopsychological approach considering the different levels of stress reactions (10, 41). The methodological implications of this approach can be characterized as a multitrait-multimethod principle (4, 26). Thus, the physiological, subjective, behavioral, and social aspects of stress should be considered as different facets of the same process. The current approach extends the traditional biopsychological stress model to a recovery-stress paradigm that allows to explain the results of burnout (13) as well as the effects of overtraining (18). In contrast to specific causes, unspecific stress processes (44), and unspecific recovery processes are considered as basic determinants of the athletes biological and psychological state. Thus, changes in the recovery-stress state are considered as an important resultant of the effects of prolonged and intensive training in high-performance athletes (25). A definition of recovery, regeneration, and mood is helpful because these concepts are used differently by various authors.

Recovery encompasses those processes of re-establishing psychological and physical resources and states which allow to tax these resources again. Recovery has physiological, subjective as well as action-oriented components. Therefore, a differentiation between physical, mood-related, emotional, behavioral, and social aspects of recovery is useful. A few examples might illustrate the different aspects of recovery. Recovery takes place on a physiological level, e.g. restoring physical resources by ingesting food, water, and minerals. Other more biological processes of recovery can be observed in the re-establishment of the physical fitness after injuries or in the humoral and biological processes during sleep (42). Of course, sleep effects psychological aspects as well. But the more psychological aspects of recovery encompass the subjective feeling of relaxation and the re-establishment of well-being and a positive mood. Behavioral aspects of recovery are those which support the biological processes as well as those which can be seen in case of recovery by changing activity from one strain to a quite different area or to leisure line activities. Social aspects of recovery can be seen in activities when people join together in groups for the weekend or for social events like having dinner with friends celebrating parties. More private and intimate aspects of social recovery concern all those situations when people have personal contact with their partners or friends.

Regeneration encloses all goal-directed processes, which allow to re-establish physical and psychological resources for performance. Some authors (1, 7) stress the active action component of regeneration. While regeneration is a kind of recovery, that is closely linked to high performance, recovery processes in general enclose more aspects. One central feature of the psychological side of regeneration and recovery is connected to positive mood and positive emotions.

Mood states are comparable to emotional states, but they are more persistent, less dynamic, and less specific than emotions (5). However, they are more transient and fluctuating than personality characteristics. Mood states, emotions, and stress should be measured on different levels, which again encompass physiological, subjective-verbal, behavioral, cognitive, and social aspects. Within emotion research, the measurement of facial expression, stature and posture turned out to be indicators of emotional states as well. Thus, mood and emotions are organismic states which are again considered as constructs with different levels of organization. Mood is more persistent and less specific than emotional states.

2. MEASUREMENT OF EMOTION, MOOD, STRESS, AND RECOVERY

As already discussed, states like emotions, mood, stress, and recovery are conceptualized as global organic states, which encompass different levels of functioning. The different components can be assessed on different organic levels (physiological, self-report, behavioral, social). However, assessment level and the indicated organic or construct level need not necessarily be the same. Measurement dimensions and the indicated aspect of the organismic states might be related in an indirect way. Using the subjective assessment approach subjects can be asked with respect to their mood and emotions, but subjects can also report on their physical state, their behavior and their social relations. On the other hand, physiological measures are not only used to indicate the physical state but to gain some objective measure of emotional reactions or the intensity of behavior. Of course, behavioral observations and performance measures are used to indicate behavioral changes, but at the same time they can be used to infer mood, emotions, physical state, and social relations. Thus, measurement level and organismic state can differ. In many cases or situations more than one measurement level is recommended, because between these different levels correlations are often fairly low. This is might be due to specificities and method specific biases (26).

2.1. The Recovery-Stress-Questionnaire

A psychometrically-based instrument to assess the recovery-stress state is the Recovery-Stress-Questionnaire (RESTQ, 12). This instrument addresses physical, subjective, behavioral, and social aspects using a self-report approach. Physical aspects are bodily symptoms associated with stress and recovery, subjective aspects are assessed by emotional mood-oriented items, behavioral aspects are mainly addressed by performance-related items and social aspects are covered by stress-related and recovery-oriented social behavior. Each of the items is formulated in an evaluated way, meaning that not the objective behavior is described, but a positively or negatively evaluated behavior. For example, Social Recovery is assessed with items like 'I had a good time with my friends' and not in a general way like 'I met people'. Each item is rated according to its frequency on a seven-point Likert-like rating scale ranging from 0 (*never*) to 6 (*always*) indicating how often the respondent participated in various activities during the *past 3 days/nights*. The first seven scales tackle different aspects of subjective strain such as General Stress, Emotional Stress, and Social Stress as well as resulting consequences. The scales Conflicts/Pressure, Fatigue, and Lack of Energy are concerned with performance aspects whereas Somatic Complaints addresses the physiological aspects of stress and strain. Success is the only resulting recovery-oriented scale which is concerned with performance in general but not in a sport specific context. Social Relaxation, Somatic Relaxation, and General Well-being are the basic scales of the recovery area with an additional scale, assessing sleep quality. Specific versions were developed for athletes (15) and coaches (16) using the modular construction of the RESTQ.

Table 1. Scales of RESTQ-Sport, the scale orientation (o), number of items (i), a sample item, and Cronbach α

Nr.	RESTQ-Sport Scales	o	i	Example	α	Retest
1	General Stress	S	4	I felt down	.75	.86
2	Emotional Stress	S	4	I was in a bad mood	.73	.82
3	Social Stress	S	4	I was angry with someone	.87	.79
4	Conflicts/Pressure	S	4	I felt under pressure	.71	.84
5	Fatigue	S	4	I was overtired	.76	.85
6	Lack of Energy	S	4	I was unable to concentrate well	.70	.86
7	Somatic Complaints	S	4	I felt uncomfortable	.73	.83
8	Success	R	4	I finished important tasks	.67	.80
9	Social Relaxation	R	4	I had a good time with my friends	.84	.91
10	Somatic Relaxation	R	4	I felt at ease	.84	.85
11	General Well-being	R	4	I was in a good mood	.85	.80
12	Sleep	R	4	I had a satisfying sleep	.84	–
13	Being in Shape	R	4	I was in a good condition physically	.89	
14	Injury	S	4	my performance drained me physically	.80	
15	Emotional Exhaustion	S	4	I felt that I wanted to quit my sport	.69	
16	Personal Accomplishment	R	4	I dealt very effectively with my teammates' problems	.79	
17	Self-regulation	R	4	I prepared myself mentally for performance	.80	
18	Disturbed breaks	S	4	my coach demanded too much of me during the breaks	.80	
19	Self-efficacy	R	4	I was convinced that I had trained well	.88	

Note. α = German Junior Rowers (n=88); S = Stress; R = Recovery; Retest = Retest-reliability for a sample of German students after 24 hours (n=72). Sleep was added to the questionnaire. Therefore, no retest-reliability data are available.

The Recovery-Stress-Questionnaire for Athletes (RESTQ-Sport, 15) consists of a basic part described above with twelve scales and additional seven sport specific scales addressing aspects of stress and recovery of athletes.[*] Sport specific aspects are Being in Shape and Injury which reflect the physical fitness of an athlete. The burnout oriented scales Emotional Exhaustion and Personal Accomplishment were developed with close resemblance to the burnout construct (29), and assess psychological symptoms of burnout (13). Self-Regulation refers to the use of psychological skills training when preparing for performance (e.g., goal setting, mental training; 43, 45). Disturbed Breaks deals with the problem of recovery deficit and disturbed recovery both of which can impair subsequent performance (21). In addition, Self-Efficacy (2) measures the level of expectation regarding an optimal performance preparation in practice.

Table 1 gives an overview of the RESTQ-Sport scales with a sample item, internal consistencies and information which scales are related to stress or recovery, respectively. Studies with German and American elite athletes have demonstrated acceptable reliability of the RESTQ-Sport scales in German (Table 1, 22) and in English [American collegiate swimmers n = 96, Cronbach's alpha ranging from .50–.88, Mdn = .78 (unpublished data)]. Table 1 includes internal consistencies of a German sample of students which have been retested after 24 hours (12).

Internal consistencies and reliabilities for the RESTQ-Sport are quite high. However, in some scales, especially as far as the somatic complaints for athletes and the per-

---

[*] The German and English versions of Recovery-Stress-Questionnaire for Athletes and for Coaches are available upon request from the authors.

formance-related scales are concerned, there are sample-dependent lower values for these coefficients. The retest-reliability leads to the conclusion that this result is probably due to the different meaning of specific items for various samples. Thus, the meaning of the scale Somatic Complaints asking items like 'I felt uncomfortable' or 'I felt physically bad' seems to be too general for athletes. These items assess Somatic Complaints of the general population, however, for athletes more sport specific oriented questions should be asked like 'I had muscle pain after performance' or 'my performance drained me physically'.The Recovery-Stress-Questionnaire was applied in different areas (12) ranging from laboratory studies (14) to various applied field settings (11, 22, 24). Interesting applications in the current context are concerned with training monitoring, with monitoring athletes' recovery-stress states during a season, and performance prediction based on the recovery-stress state on an individual basis (13, 20).

2.2. The Profile of Mood States

An instrument that has repeatedly been used to measure with 65 items mood states of athletes is the Profile of Mood States (POMS, 31) which assess transient, fluctuating affective states including tension, depression, anger, fatigue, confusion, and vigor. Each mood dimension of the English version is rated on a scale of 1 to 4 from 'not at all' to 'extremely'.

In contrast to the original, the German translation of the POMS (32) consists only of the scales Depression, Anger, Fatigue, and Vigor. In the German version adjectives are rated on a scale from 1 to 7 'not at all' to 'very strongly'.* The reduction down to 35 items and four scales seems to be due to the seven-point rating scale (compared to the four-point rating scale in the original), the translation of the questionnaire, and the subsequent psychometric testing. Factor analysis revealed just the four scales mentioned above (32).

2.3. The Relationship between the RESTQ-Sport and the POMS

Two studies dealing with the RESTQ-Sport and the POMS should be summarized. To match the reference time of the RESTQ-Sport, the time frame in the heading of the POMS was changed in both studies to 'in the past three days/nights'.

2.3.1. Study I. American female collegiate swimmers (n = 30) completed voluntarily the RESTQ-Sport and the POMS three times during the course of the season (19). The first meet occurred six days prior to the four day Thanksgiving break while the second took place two days after the break. These two meets were chosen because it was assumed that the recovery-stress state might differ prior and following the break. The third competition took place three months later before the Southeastern Conference Championships. The swimmers were instructed to complete the questionnaires 48 hours prior to each of the three intercollegiate competitions. Due to incomplete data for single events the number of swimmers for the following analysis was reduced down to $n = 18$ subjects (age: $M = 19.67$ yr, $SD = 1.18$). The correlational analysis of this data set showed that only the scale Vigor appears to be positively related to the recovery scales of the RESTQ-Sport (e.g., Social Relaxation $r = .59$; Somatic Relaxation $r = .54$; General Well-being $r = .60$; Sleep $r =$

---

* In the German adaptation of the POMS scales are labeled Niedergeschlagenheit (Depression), Mißmut (Anger), Müdigkeit (Fatigue), and Tatendrang (Vigor). The adjectives are rated on a scale from 'überhaupt nicht (not at all)' to 'sehr stark (very strongly)'.

.48), while Tension, Depression, Anger, Fatigue, and Confusion are negatively correlated with recovery. Vice versa, a positive relationship exists between the stress-related scales of the RESTQ-Sport and Tension, Depression, Anger (with e.g., Emotional Stress r = .74), Fatigue, and Confusion whereas Vigor seems to be negatively correlated with stress (17). Kellmann et al. (19) could show that the recovery-stress state of the athletes changed significantly. The stress level diminished and the recovery level increased over the holiday break. This result, of course, is no surprise since 'taking a break' is the purpose of a vacation. However, measuring this effect taking behavioral oriented activities into consideration is quite unique. It is also important to look at individual situations. Athletes who had to travel a long way to their homes could not necessarily profit at the same level from the break as swimmers who lived nearby the university. For some athletes Thanksgiving holidays was a time to rest and enjoy a slower pace of life. For others, it was characterized by a higher level of emotional intensity associated with activities involving family, friend and significant others. While such activities may represent a change from the normal training regimen, they may be no less emotionally exacting.

Since the interrelation between Emotional Stress and Anger is quite high (r = .74) the variation across the three competitions will be shown in Figure 1. The upper panel demonstrates the change for Emotional Stress measured by the RESTQ-Sport, the lower panel for Anger assessed by the POMS.

Changes across time were analyzed using trend parameters which were obtained by a data transformation with orthogonal polynomials (8). The development Emotional Stress and Anger corresponds for the three competitions. Orthogonal polynomials reveal a significant overall test ($F_{(2,16)}=8.53$; $p < 0.01$) with a quadratic trend for the dependent variable Emotional Stress ($F_{(1,17)}=18.13$; $p < 0.01$). A decline from the first assessment to the second competition which took place after the four day Thanksgiving break was followed by an increase to the third event. The analysis for Anger [overall test ($F_{(2,16)}=7.17$; $p < 0.01$), quadratic trend ($F_{(1,17)}=15.25$; $p < 0.01$) underlines statistically the corresponding change.

Similar results were found for Somatic Complaints (RESTQ-Sport) and Fatigue (POMS). Figure 2 shows the alteration for both scales. For Somatic Complaints the change across time and the trend components turned out to be significant [overall test ($F_{(2,16)}=26.7$; $p < 0.001$), linear trend ($F_{(1,17)}=15.97$; $p < 0.01$), quadratic trend ($F_{(1,17)}=22.48$; $p = 0.001$). This also occurred for Fatigue [overall test ($F_{(2,16)}=14.94$; $p < 0.001$), linear trend ($F_{(1,17)}=17.2$; $p < 0.01$), quadratic trend ($F_{(1,17)}=27.92$; $p = 0.001$). Both questionnaires seem to be sensitive for events in the life of athletes that effects the recovery-stress state and mood, respectively.

2.3.2. Study II. The German Junior Rowing National Team (n = 58; age: M = 17.1 yr, SD = .72) completed the German version of the POMS and of the RESTQ-Sport overall six times during preparation camp for the 1998 World Championships in Austria. Table 2 gives an overview of the correlations between the RESTQ-Sport and the POMS at the second measurement.

Although the RESTQ-Sport and the POMS apply different types of scales—frequency vs. intensity (6)—analysis revealed close and theoretically expected correlational patterns which matches the results already presented by Kellmann (17). Depression, Fatigue and Anger are negatively correlated with recovery-related scales while for Vigor a positive relationship occurs. The stress-related RESTQ-Sport scales show a positive interrelation with Depression, Fatigue, and Anger but a negative with Vigor. This correlational pattern is stable for the other measurements. The specificity of the relationships between these instruments acquires special attention, and therefore, selected correlations will be

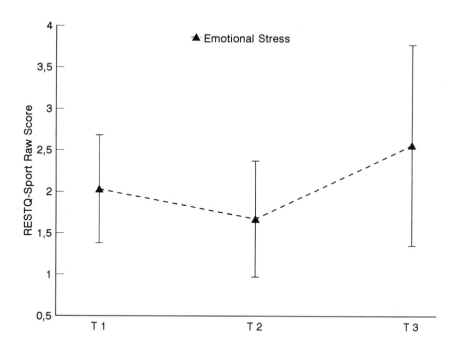

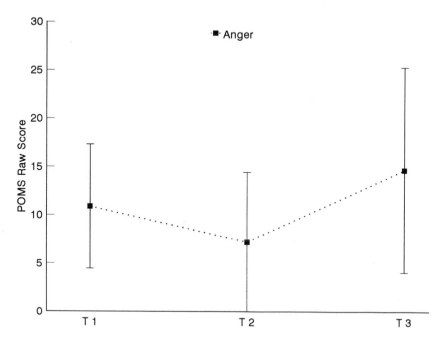

Figure 1. Means and standard deviations (SD) of the RESTQ-Sport scale Emotional Stress (upper panel, broken line) and the POMS scale Anger (lower panel, dotted line).

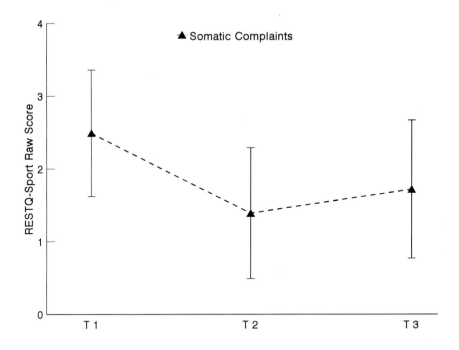

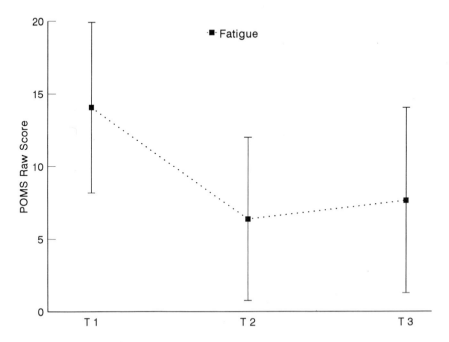

Figure 2. Means and standard deviations (SD) of the RESTQ-Sport scale Somatic Complaints (upper panel, broken line) and the POMS scale Fatigue (lower panel, dotted line).

Table 2. Correlations of RESTQ-Sport and the POMS

RESTQ-Sport Scales	POMS Scales				
	Depression	Fatigue	Vigor	Anger	Retest
1. General Stress	.69**	.52**	−.61**	.56**	.71**
2. Emotional Stress	.68**	.45**	−.52**	.68**	.72**
3. Social Stress	.43**	.31*	−.34*	.67**	.77**
4. Conflicts/Pressure	.58**	.27	−.26	.47**	.73**
5. Fatigue	.25	.74**	−.57**	.28	.81**
6. Lack of Energy	.53**	.62**	−.53**	.37*	.68**
7. Somatic Complaints	.35*	.70**	−.56**	.33*	.76**
8. Success	−.17	−.16	.41**	−.08	.70**
9. Social Relaxation	−.31*	−.22	.49**	−.22	.74**
10. Somatic Relaxation	−.34*	−.54**	.73**	−.28	.79**
11. General Well-being	−.52**	−.40*	.66**	−.38*	.61**
12. Sleep	−.43**	−.38*	.46**	−.41**	.70**
13. Fitness/Being in Shape	−.33*	−.64**	.73**	−.26	.71**
14. Fitness/Injury	.21	.51**	−.35*	.20	.59**
15. Emotional Exhaustion	.68**	.65**	−.57**	.54**	.72**
16. Personal Accomplishment	−.40*	−.21	.34*	−.13	.81**
17. Self-Regulation	−.3300*	−.26	.55**	−.14	.77**
18. Disturbed Breaks	.23	.34*	−.56**	.32*	.64**
19. Self-Efficacy	−.45**	−.27	.54**	−.30	.82**
Retest	.83**	.82**	.78**	.65**	

Note. Correlations marked with * are significant on the $p < .01$ level (one-tailed). Correlations marked with ** are significant on the $p < .001$ level (one-tailed). Retest = Retest-reliability after three days. Sample: German Junior Rowing National Team of 1998 (n=58).

discussed. POMS-Depression correlates with General Stress and Emotional Stress as well as with Emotional Exhaustion which is associated with the burnout syndrome (29). POMS-Fatigue shows high positive correlations with somatic oriented scales of the RESTQ-Sport such as Fatigue, Lack of Energy, Somatic Complaints, and negative for Fitness/Being in Shape. In addition the relationship to Emotional Exhaustion turned out significant. The pattern of POMS-Vigor almost mirrors that of POMS-Fatigue. Vigor correlates positively with the recovery-related scales Somatic Relaxation, General Well-being, and Fitness/Being in Shape, and negatively with General Stress. POMS-Anger correlates just with Social Stress and Conflicts/Pressure above $r > .6$. Thus, there are correlational data, but the conceptual and theoretical relationships between emotions, mood, and the recovery-stress state need further investigations to understand the underlying processes. Table 2 lists the retest-reliability of the RESTQ-Sport and the POMS after three days, ranging for the POMS between $r = .55$ and $r = .68$. For the RESTQ-Sport, the range of retest-reliability is slightly broader lying between $r = .46$ (Disturbed Breaks) and $r = .76$ (Self-Regulation). Taken together, beside the correlations for Disturbed Breaks and Self-Regulation the retest-reliability of both questionnaires seems to be quite similar.

2.4. Training Monitoring Using the Recovery-Stress-Questionnaire for Athletes

As can be seen from Figure 3, the recovery-stress state changes with training load in a sample of the German Junior Rowing National Team during preparation camp for the 1995 World Championships. Training load has been operationalized by rowed kilometers

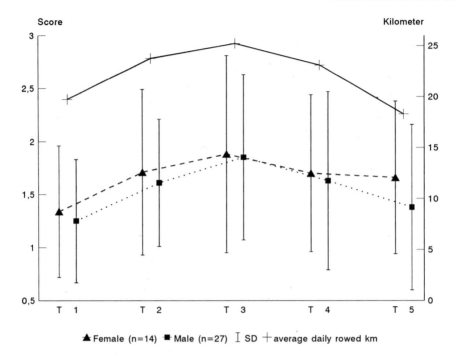

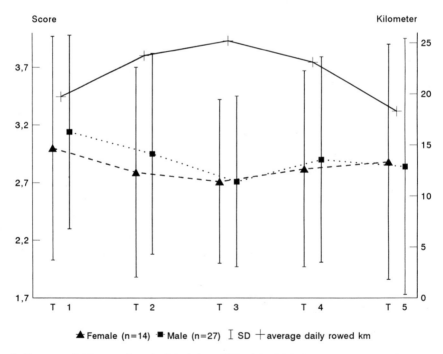

Figure 3. Upper panel: Means and standard deviations (SD) of the RESTQ-Sport scale Lack of Energy for females (broken line), for males (dotted line), and the average number of daily rowed kilometers (solid line) throughout training camp. Lower panel: Means and standard deviations of the RESTQ-Sport scale Somatic Relaxation for females (broken line), for males (dotted line), and the average number of daily rowed kilometers (solid line) throughout training camp.

per day. Female as well as male athletes showed that changes in the RESTQ-Sport scale Lack of Energy (upper panel) closely correspond to the training load (22). The lower panel demonstrates that Somatic Relaxation is invertedly related to the training load during the training camp.

The present results suggest that a *dose-response relationship* exists between training volume, indicated by the average number of daily rowed kilometers, and the subjective assessment of *somatic* components of stress and recovery. High duration is indicated by elevated levels of stress and simultaneous lowered levels of recovery. It is in line with the results of Morgan and his colleagues (34, 38, 40) who found that increases in training volume parallel corresponding elevations in mood disturbance, and mood improvements occur if training is reduced.

2.5. Individual Assessment of the Recovery-Stress State

Applied researchers focus on the concept of the athletes' recovery-stress state during preparation camp and before preliminaries. For individual assessment an 'area of tolerance' is calculated by the mean ± standard deviation of the whole rowing team. In this area the score in considered to be completely in range. If more than two scores deviate they receive particular attention if specific patterns occur. However, since training volume changes throughout training camp (see Figure 3) the 'area of tolerance' for the recovery-stress state has to be calculated separately for each measurement as a norm for the individual assessment at the specific time of the training process. Using this approach the change of individual profiles over time can be interpreted as well as the individual profile in reference to the team. To illustrate the applied use of the RESTQ-Sport, the case of a female rower of the German Junior Rowing National Team (22) should briefly be described. Her recovery-stress state revealed very low scores in the scale Sleep which provides a combined pattern with the low values in Somatic Relaxation, and elevated scores in Fatigue (Figure 4).

The coach had no explanation for the results but found a simple solution when he talked to the athlete. She was sleeping in a supplemental bed which was in very bad shape. After putting a board below the mattress, she slept better and her scores improved throughout the next measurements. The solution is not always that simple but often easier then expected. In this case, the athlete did not talk to the coach at all about what bothered her but expressed the current situation using the RESTQ-Sport.

2.6. Monitoring the Recovery-Stress State during a Worldcup Season with Mountainbikers

During the Mountainbike Worldcup in 1997, a group of 17 athletes filled out the RESTQ-Sport on a week by week schedule. Data were collected in the course of twelve weeks during the season. Data from all mountainbikers and all measurement occasions were pooled to obtain an average score and an estimation of the standard deviation. Based on these data, a score was computed that possibly reflects the recovery-stress state for athletes quite well. The scores of stress-related scales (scale 1 to 7, 14, 15, and 18; see Table 1) were summed up and divided by the number of scales representing the 'stress score'. The same procedure was used for the recovery-oriented scales (scale 8 to 12, 13, 16, 17, and 19; see Table 1) resulting in a 'recovery score'. The 'stress score' as well as the 'recovery score' were converted to standardized values (z-values) by subtracting the global sample mean and dividing the difference by the standard deviation. Thus, a standardized recovery and a stress score could be obtained on a common scale, which allows to

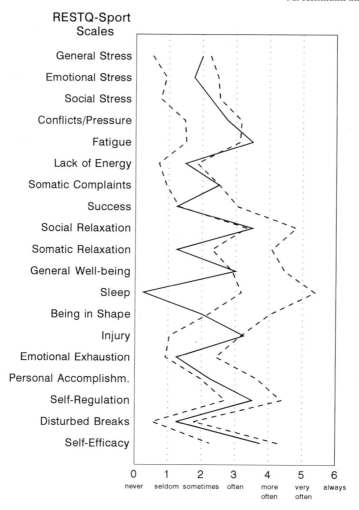

Figure 4. Recovery-stress state for a female rower (solid line) and the reference norm of the whole team (mean ± one standard deviation; between the broken lines).

compute a difference between stress and recovery. This score can be interpreted as a kind of athletes' resource measure. Two single-case studies will be given in the next figures.

Figure 5 shows the large variations in the recovery-stress state of this biker during the season, with a peak in stress and a down in recovery during the mid of July. That was inverted for the end of August and broke down again by the beginning of September. Interestingly, this biker was diagnosed as suffering from an overtraining syndrome before the race on the 24th of July.

The biker with the code number 15 (Figure 6) shows a quite different pattern of recovery and stress scores. There is a peak in recovery in the mid of July and the stress scores are below average starting with the 10th and 17th of July. Thus, we can see a kind of positive resource for this athlete during the time from the 24th to the 31st of July. The decline in stress and the increase in recovery during the mid of July can be interpreted as a tapering phase that enabled him to win the European Championships. By the end of the season, the pattern for this biker turned around. The recovery-stress state indicates that re-

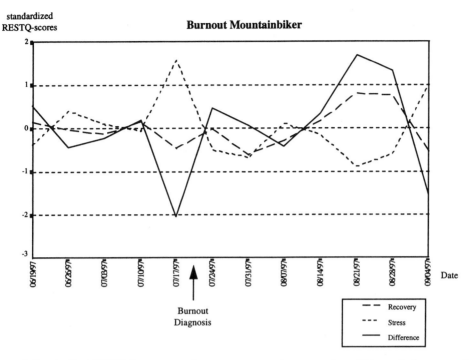

Figure 5. Standardized RESTQ-Sport scores of recovery, stress, and recovery-stress difference in the course of a season for an overtrained athlete.

covery peaks negatively and stress peaks positively which might be interpreted in the sense that the available resources are worn out.

3. DISCUSSION

The results presented in this chapter show that on a group and on an individual level, the recovery-stress state seems to be an important psychological variable which is associated with the training load of an athlete. Thus, the recovery-stress state as well as the mood state allow to have a very economic and easy tool to monitor athletes' training processes during preparation camps and before regular competitions within a season. It has to be considered, that the RESTQ-Sport and the POMS are not direct measures of physiological states of the organism, both instruments reflect the subjective representation of these states.There are different possibilities how the recovery-stress state are related to mood and overtraining. The recovery-stress state of an athlete can just be correlated to mood, due to something like an intervening third variable, like training load. For example, a change in the athletes' metabolism might affect the recovery-stress state and mood. On the other hand, it might also be seen that the recovery-stress state of the athlete is a very important determinant of the athlete's mood. Laboratory experiments showed that stress as well as a simple recovery activity can be used to change the emotional state of a person drastically (14). Thus, there is some evidence, that the recovery-stress state affects mood. It also has to be considered that mood might affect the scoring the recovery-stress state or the way the questionnaire is answered. It seems to be very

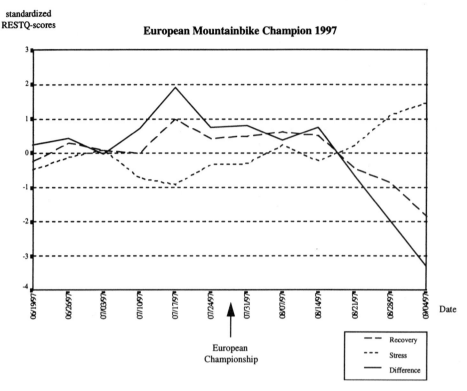

Figure 6. Standardized RESTQ-Sport scores of recovery, stress, and recovery-stress difference in the course of a season with optimal preparation for a Championship.

likely that stress-recovery state and mood are interdependent organismic states. The data presented in this chapter support that the RESTQ-Sport and the POMS are valuable tools to prevent overtraining. Similarities were found in the *dose-response relationship* (training volume/recovery-stress state, mood), a close interrelation of the scales, and even the retest-reliability seems to be in a comparable range. Consequently the question arises *'What is the advantage of the RESTQ-Sport'*? The advantage of the Recovery-Stress-Questionnaire for Athletes compared with the Profile of Mood States results from the systematic multi-dimensional approach. While for the POMS we have the 'Iceberg-Profile', primarily consisting of negative mood states and only one aspect covering positive mood states, the Recovery-Stress-Questionnaire allows to have a very differently shaped picture of the athletes' state without using much more time to assess the state. From the actual recovery-stress profile (Figure 4) one can derive concrete solutions to actual problems and of course this profile might be used to derive specific intervention strategies (22). The RESTQ-Sport provides coaches, sportpsychologists, and athletes with important information during the training process. This of course also applies when coaches and athletes have not been in contact for a longer period of time (e.g., caused by vacation or injury). Receiving information about the actual state based on information from the past days is important to start with the training process on an adequate level meaning not to overtrain an athlete but giving an adequate training stimulus. In addition, the perception of coaches and athletes may differ, and therefore, the coaches get more information about the subjective perception of an athlete. In this context, the practical experience with the Junior National Rowing Team shows that athletes use the RESTQ-Sport 'to express themselves'. This can be supported by the fact that, before important competitions, athletes become more sensitive to certain activities, and perceive the environment differently, although the objective environment and the coaches view does not change. Using this questionnaire, athletes and coaches could become aware of the importance of daily activities, how these events are related to stress/recovery and about their impact on performance in sport. In addition, the module concept of the RESTQ-Sport divides the questionnaire in a 'basic part' (first twelve scales) dealing with the general life outside of sports (12). The other scales refer to a sport specific context with sportpsychological oriented areas such as Self-Regulation or Self-Efficacy which indeed becomes more important as a competition approaches. Another possibility provided by the RESTQ-Sport is the observation of individual and/or group processes over a longer time period (22). The assessment up to 48 hours before competition allows enough time to intervene for an optimal initial state by the coach or sportpsychologist. An important function of emotions and mood is serving as an organismic signal system. Mood states and emotions inform us about our organismic state and appraise our person-environment interactions. Thus, instruments which assess mood state or the recovery-stress state serve as efficient indicators about the athletes' organismic and psychological state. Based on this, the RESTQ-Sport predicts performance (13, 20, 23), possibly because the athletes can report on their state by using this instrument. The RESTQ-Sport can therefore be used as a kind of 'mirror' showing the coaches as well as the athletes some important aspects of their personal state. From the view of Hanin's 'individual zones of the optimal functioning' model (8), this is also an important information. The main difference, compared with Hanin's model, is, that the zones of functioning should be assessed individually and not on a group level to obtain a performance prediction. As the RESTQ-Sport allows a prediction on the basis of group data, one can possibly combine the individual standardization and the group standardization approach in case that no long-term observation is possible for the individual athlete. Otherwise, the recovery-stress state might well be used to assess the 'individual zones of optimal functioning'. It is highly recommended to use standardized psychometrically-tested instruments from research in psychology when assessing or monitoring psychological states of athletes. On the other hand, the psychological variables which are, of course, sensitive to non-sports specific changes in athletes' states, have to be supplemented by sportmedical assessments of the athletes' state. Using these different measurement levels, it seems possible to improve the prediction of overtraining and to differentiate between overreaching and the induction of staleness.

4. CONCLUSIONS

Using a biopsychological stress model to describe and explain the changes with (over-) training, the recovery-stress state of athletes turned out to be a central psychological construct of training monitoring. This concept is closer and more directly related to overtraining and performance than mood and emotions as it is closely linked to the athletes' daily activities. A second implication of the stress model is methodological in nature: A multidimensional interdisciplinary approach of sportpsychology, sportmedicine, and coaching science is recommended to provide a better understanding of overreaching and top-performance, as well as overtraining, staleness, and burnout.

REFERENCES

1. Allmer H (1996) Erholung und Gesundheit: Grundlagen, Ergebnisse und Maßnahmen [Recovery and health: basics, results and interventions]. Hogrefe Göttingen
2. Bandura A (1977) Self-efficacy: Toward a unifying theory of behavioral change. Psychological Review 84:191–215
3. Borg G (1975) Perceived exertion as an indicator of somatic stress. Scandinavian Journal of Rehabilitational Medicine 2:92–98
4. Campbell A, Fiske DW (1956) Convergent and discriminant validation by the mulititrait-multimethod matrix. Psychological Bulletin 56:276–279
5. Carver CS, Scheier MF (1990) Origins and functions of positive and negative affect: A control-process view. Psychological Review 97:19–35
6. Diener E, Emmons R (1984). The interpedendence of positive and negative affect. Journal of Personality and Social Psychology 47:105–117
7. Eberspächer H, Renzland J (1988) Regeneration im Sport [Regeneration in sports]. bps Köln
8. Ferguson GA, Takane Y (1989) Statistical Analysis in Psychology and Education. McGraw-Hill New York
9. Hanin Y (1997) Emotions and athletic performance: Individual zones of optimal functioning model. European Yearbook of Sports Psychology 1:29–72
10. Janke W, Wolffgramm J (1995) Biopsychologie von Streß und emotionalen Reaktionen: Ansätze interdisziplinärer Kooperation von Psychologie, Biologie und Medizin [Biopsychology of stress and emotional reactions: Starting points of an interdisciplinary cooperation of psychology, biology, and medicine]. In: Debus G, Erdmann G, Kallus KW (eds) Biopsychologie von Streß und emotionalen Reaktionen. Hogrefe Göttingen:293–349
11. Johnson MS, Wrisberg CA, Kellmann M, Kallus KW (1997) Assessing Stress and Recovery in Collegiate Athletes. Journal of Applied Sport Psychology (Supplement) 8:S113
12. Kallus KW (1995) Der Erholungs-Belastungs-Fragebogen (EBF) [The Recovery-Stress-Questionnaire]. Swets & Zeitlinger Frankfurt
13. Kallus KW, Kellmann M Burnout in sports: A recovery-stress state perspective. In: Hanin Y (ed) Emotion in Sport. Human Kinetics Champaign, in press
14. Kallus KW, Krauth J (1995) Nichtparametrische Verfahren zum Nachweis emotionaler Reaktionen [Nonparametric methods for the provement of emotional reactions]. In: Debus G, Erdmann G, Kallus KW (eds) Biopsychologie von Streß und emotionalen Reaktionen. Hogrefe Göttingen:23–43
15. Kallus KW, Kellmann M (1994) The Recovery-Stress-Questionnaire for Athletes [unpublished questionnaire]. Würzburg University Würzburg
16. Kallus KW, Kellmann M (1995) The Recovery-Stress-Questionnaire for Coaches. In: Vanfraechem-Raway R, Vanden Auweele Y (eds) Proceedings of the IXth European Congress on Sport Psychology in Brussels. FEPSAC/Belgian Federation of Sport Psychology Brussels:26–33
17. Kellmann M (in press). Die Beziehungen zwischen dem Erholungs-Belastungs-Fragebogen für Sportler und dem Profile of Mood States [The relationship of the Recovery-Stress-Questionnaire for Athletes and the Profile of Mood States]. In: Alfermann D, Stoll O (eds) Motivation und Volition im Sport - Vom Planen zum Handeln. bps Köln
18. Kellmann M (1997) Die Wettkampfpause als integraler Bestandteil der Leistungsoptimierung im Sport: Eine empirische psychologische Analyse [The Rest Period as an Integral Part for Optimizing Performance in Sports: An Empirical Psychological Analysis]. Dr. Kovac Hamburg
19. Kellmann M, Johnson MS, Wrisberg CA (1998). Auswirkungen der Erholungs-Beanspruchungs-Bilanz auf die Wettkampfleistung von amerikanischen Schwimmerinnen [Effects of the recovery-stress state on performance of American female swimmers. In: Teipel D, Kemper R, Heinemann D (eds) Sportpsychologische Diagnostik, Prognostik und Intervention. bps Köln:123–126
20. Kellmann M, Kallus KW (1993) The Recovery-Stress-Questionnaire: A potential tool to predict performance in sports. In: Nitsch JR, Seiler R (eds) Movement and Sport: Psychological Foundations and Effects. Academia Sankt Augustin:242–247
21. Kellmann M, Kallus KW (1994) Interrelation between stress and coaches' behavior during rest periods. Perceptual and Motor Skills 79:207–210
22. Kellmann M, Kallus KW, Günther KD, Lormes W, Steinacker JM (1997) Psychologische Betreuung der Junioren-Nationalmannschaft des Deutschen Ruderverbandes [Psychological consultation of the German Junior National Rowing Team]. Psychologie und Sport 4:123–134
23. Kellmann M, Kallus KW, Kurz H (1996) Performance Prediction by the Recovery-Stress-Questionnaire. Journal of Applied Sport Psychology (Supplement) 8:S22

24. Kellmann M, Kallus KW, Steinacker J, Lormes W (1997) Monitoring stress and recovery during the training camp for the Junior World Championships in Rowing. Journal of Applied Sport Psychology (Supplement) 9:S114
25. Kreider RB, Fry AC, O'Toole ML (eds) (1998) Overtraining in sports. Human Kinetics Champaign
26. Laux L (1983) Psychologische Streßkonzeptionen [Psychological conceptions of stress]. In: Thomae H (ed) Enzyklopädie der Psychologie. Hogrefe Göttingen:453–535
27. Lazarus RS, Launier R (1978) Stress-related transactions between person and environment. In: Perwin LA, Lewis M (eds) Perspectives in international psychology. Plenum New York:287–327
28. Lehmann M, Foster C, Keul J (1993) Overtraining in endurance athletes: A brief review. Medicine and Science in Sports and Exercise 25:854–862
29. Maslach C, Jackson SE (1986) Maslach Burnout Inventory. Consulting Psychologists Press Palo Alto
30. Meyers AW, Whelan JP (1998) A systemic model of understanding psychological influences of overtraining. In: Kreider RB, Fry AC, O'Toole, ML (eds) Overtraining in sport. Human Kinetics Champaign:335–369
31. McNair DN, Lorr M, Droppleman LF (1971, 1992) Profile of Mood States. Manual. Educational and Industrial Testing Service San Diego
32. McNair DN, Lorr M, Droppleman LF, Biehl B, Dangel S (1981) Profile of Mood States (German adaptation). In: Collegium Internationale Psychiatriae Scalarum (CIPS). Beltz Weinheim:without page number
33. Morgan WP (1985) Selected psychological factors limiting performance: A mental health model. In: Clarke DH, Eckert HM (eds) Limits of human performance. Human Kinetics Champaign:70–80
34. Morgan WP, Brown DR, Raglin JS, O'Conner PJ, Ellickson KA (1987) Psychological monitoring of overtraining and staleness. British Journal of Sport Medicine 21:107–114.
35. Morgan WP, Costill DL (1996) Selected psychological characteristics and health behaviors of aging marathon runners: A longitudinal study. International Journal of Sport Medicine 17:305–312
36. Morgan WP, Costill DL, Flynn, MG, Raglin JS, O'Conner P (1988) Mood disturbance following increased training in swimmers. Medicine and Science in Sports and Exercise 20:408–414
37. Noble BJ, Robertson RJ (1996) Perceived exertion. Human Kinetics Champaign
38. O'Conner PJ, Morgan WP, Raglin JS (1991) Psychobiologic effects of 3d of increased training in female and male swimmers. Medicine and Science in Sports and Exercise 23:1055–1061
39. Raglin JS (1993) Overtraining and staleness: Psychometric monitoring of endurance athletes. In: Singer RB, Murphey M, Tennant LK (eds) Handbook of research on sports psychology. Macmillan New York:840–850
40. Raglin JS, Morgan WP, O'Conner PJ (1991) Changes in mood state responses during training in female and male college swimmers. International Journal of Sports Medicine 12:585–589
41. Stanford SC, Salomon P (eds) (1993) Stress. From synapse to syndrome. Academic Press London
42. Savis J (1994) Sleep and athletic performance: Overview and implications for sportpsychology. The Sport Psychologist 8:111–125
43. Seiler R (1992) Performance enhancement - A psychological approach. Sport Science Review 1:29–45
44. Selye H (1993) History of the stress concept. In: Goldberger L, Breznitz S (eds) Handbook of stress. The Free Press New York:7–17
45. Vealey R (1988) Future directions in psychological skill training. The Sport Psychologist 2:318–336

9

THE EXERCISE MYOPATHY

Teet Seene, Maria Umnova, and Priit Kaasik

University of Tartu
Tartu EE2400, Estonia

1. INTRODUCTION

The modern training of athletes is based on the overload principle and negative feed-back theory: the training stimulus must be strong enough to induce disturbance of homeo-stasis so that the body has to initiate reactions to adapt to the training stimulus (7). This means that overloading is a natural part of an athletes training process and provides stim-uli for adaptation and supercompensation. Imbalance in the training load–recovery rela-tionship is the primary factor contributing to overtraining syndrome (2,6). The reflection of overtraining syndrome on the ultrastructural level of skeletal muscle has revealed that the damaging effect depends on the muscle fiber type (13). Due to the destruction of myofibrils and atrophy of muscle fibers, exercise myopathy develops as a result of over-training (8, 13). The most sensitive to the long-lasting exhaustive endurance exercise are fast-twitch muscle fibers (11). The aim of the present study was to investigate the reflec-tion of exercise myopathy on the ultrastructural level of fast-twitch muscle fibers.

2. METHODS

Male rats of the Wistar strain, 16–17 weaks old were maintained on a constant diet SDS-RM 1 (C) 3/8 (SDS, Withaus, Essex, England). Food and water were given *ad libi-tum*. The rats were housed four per cage in plastic cages at 12/12 h light/dark period. The experiments were approved by the ethical commission on the Medical Faculty of Tartu University, licence No. 193. The overtraining was carried out as described previously (8). For studies of the fast-glycolytic (type II B) and fast-oxidative-glycolytic (type II A) fibers the *m. quadriceps femoris* dissected and cytochromes aa_3 and myoglobin and Myosin Heavy Chain isoforms were used as markers for type of muscle fibers (5, 10). At the ul-trastructual level the types of muscle fibers were determined on the basis of their morpho-logical differences (11). Ultrathin sections were cut from transversely and longitudinally oriented muscle blocks, stained with uranyl acetate and lead citrate (11).

Overload, Performance Incompetence, and Regeneration in Sport, edited by Lehmann *et al.*
Kluwer Academic / Plenum Publishers, New York, 1999.

3. REFLECTION OF EXERCISE MYOPATHY ON THE ULTRASTRUCTURE OF OXIDATIVE-GLYCOLYTIC MUSCLE FIBERS

We have established both specific and non-specific changes in the myopathic fast-twitch muscle fibers and changes which appear in skeletal muscles in other stressful conditions. Non-specific changes include the appearence of myelin figures in complexes of mitochondria, absence of mitochondria and formation of their gigantic forms. We regard as specific changes the destruction of peripheral myofilaments, attenuation of myofibrils and complete destruction of some sarcomeres (Figure 1).

Due to the destruction of myofibrils the sarcoplasm-filled spaces between myofibrils increase. These spaces contain mitochondria in which can be seen a quantity of cristae and a compact matrix, fragments of T-tubules, fragments of the sarcoplasmic reticulum and a quantity of glycogen granules. In some myofibrils warped Z-lines can be observed (Figure 2).

The destruction of muscle fiber organelles is accompanied by the activation of lysosomal structures. Long-term muscle exertion causes a decrease in the osmosis of lipid inclusions which is most probably connected with the more intensive use of fatty acids as substrate of energy (Figure 3).

In the fast-oxidative-glycolytic muscle fibers there occur also some satellite cells. Under the light microscope there emerges a clear tendency towards increasing the number of nuclei in these muscle fibers that contain 2–3 nucleoli. The electron microscopic studies confirm this. More extensive and more clearly expressed destructive changes occur in the myopathic fast-oxidative-gylcolytic muscle fibers bringing about atrophy of those muscle fibers. At the same time those muscle fibers reveal morphological symptoms of the

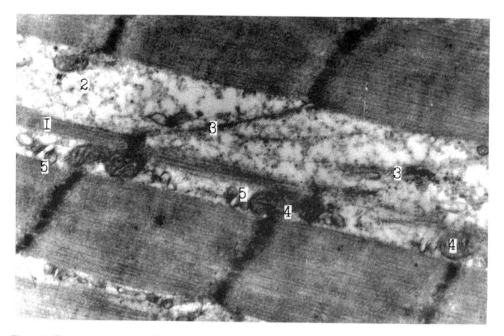

Figure 1. Electron micrograph of fragment of fast-oxidative-glycolytic muscle fiber in the state of contraction. Almost all of the myofibril is destructed (1) and in the result wide intermyofibrillar space has formed (2). Irregularly located thick myofilaments in the sarcoplasma (3), mitochondria (4), triads (5). Magnification, x 32000.

Figure 2. Electron micrograph of a fragment of fast-oxidative-glycolytic muscle fiber. Destructions of myofibrils (1), thin myofibrils (2), damage of the Z-line (3), a quantity of glycogen granules in the wide intermyofibrillar space (4), mitochondria on both sides of Z-line (5), T-tubules on the level of the border of A and I discs (6). Magnification, x 27000.

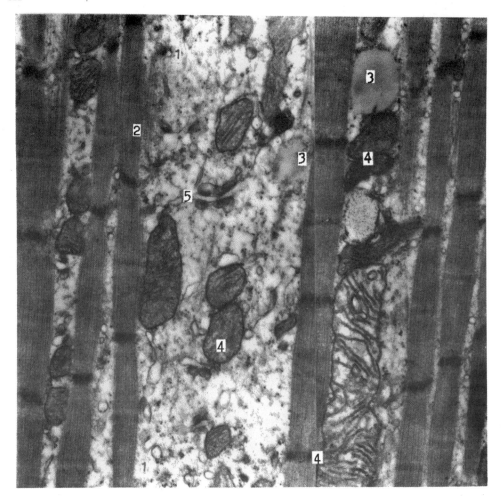

Figure 3. Electron micrograph of fragment of fast-oxidative-glycolytic muscle fiber. Wide intermyofibrillar space due to destruction of myofibrils (1), thin myofibrils (2), lipid droplets (3), located among mitochondria (4), and T-tubules (5). Magnification, x 26000.

continuation of restoration processes: increase in the number of nucleoli, satellite cells under the basal membrane, activation of the mitochondrial system, the occurrence of numerous lipid drops and glycogen granules.

3.1. Ultrastructure of the Neuromuscular Junctions in the Myopathic Fast-Oxidative-Glycolytic Muscle Fibers

The appearance in the lysosomes of the fast-oxidative-glycolytic muscle fibers synaptic terminal of multivesicular formations and large multi-layered structures which consisting of several parallel membranes and containing besides vesicles a quantity of axoplasm (Figure 4), is typical for the myopathic muscle.

In the terminals there occur also such multimembrane formations that have developed from thickened mitochondria, and there also occur both small and large synaptic vesicles. Most of the vesicles are concentrated in the central part of the terminal and only

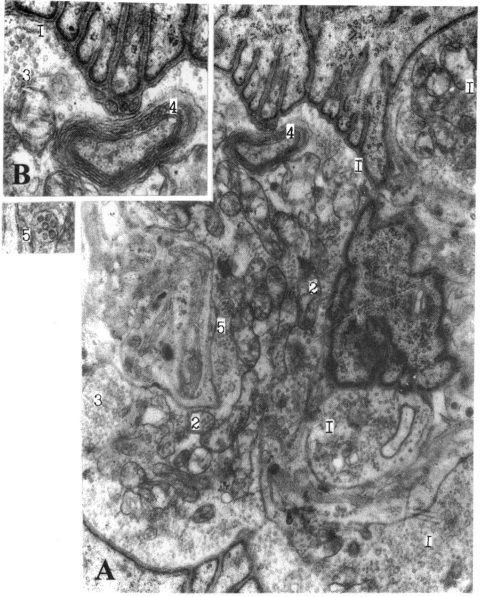

Figure 4. Electron micrograph of neuromuscular junction on the fast-oxidative-glycolytic muscle fiber. Branching of the axoral terminal (1), a quantity of mitochondria (2), synaptic vesicles (3), multimembrane structure (4) and multivesicular body (5). Magnification, A-x 14000, B-x 33000.

single groups of vesicles can be found in the active zone of the terminals—in those regions of the presynaptic membrane which correspond to the postsynaptic folds (Figure 5).

It seldom occurs that coated vesicles are found here. There are plenty of mitochondria in those terminals which contain very few cristae. The axon terminals branch off fairly often, and this leaves the impression that their range is quite large. There also occur regions with a widened synaptic slot, regions which contain membranous structures and vesicles. It is evident that in these regions the nerve-muscle transfer is disrupted (Figure 6).

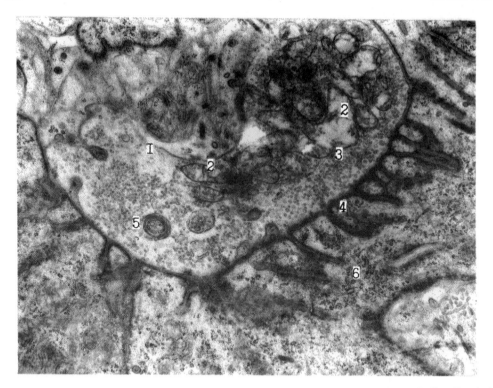

Figure 5. Electron micrograph of neuromuscular junction on the fast-oxidative-glycolytic muscle fiber. Terminal (1) is filled with a great number of mitochondria (2) and synaptic vesicles (3). Groups of synaptic vesicles located near the perisynaptic membrane on the level of post-synaptic folds (4); multimembrane structure (5); ribosomes (6) in the postsynaptic region. Magnification, x 14000.

Around the terminals there form several postsynaptic folds which branch off quite often. The numerous mitochondria located in the postsynaptic region are tightly packed with cristae. Between the mitochondria lysosome-like formations can be found. In the postsynaptic region there occur some nuclei of the muscle cell, glycogen granules, ribosomes and canals of the granular reticulum. The existence of the granular reticulum points to the occurrence of protein synthesis in the region.

When comparing structural changes in the nerve-muscle synapses of fast-oxidative-glycolytic muscle fibers in case endurance training and in overtraining caused myopathy, one has the impression that in the first case the changes are smaller in scope. And this is why the structural heterogeneity of the synapse is not so clearly manifest. In myopathic muscle the synapse contains several vesicles, there are fewer coated vesicles and the terminals contain fewer mitochondria. In this case we were not able to find satellite cells in the postsynaptic region.

4. REFLECTION OF EXERCISE MYOPATHY ON THE ULTRASTRUCTURE OF FAST-GLYCOLYTIC MUSCLE FIBERS

In myopathic muscle the myofibrils in glycolytic muscle fibers are damaged. Most damage is done to the myofibrils in the periphery of the muscle fibers although in comparison with the analogous destruction occurring in the oxidative-glycolytic fibers it is

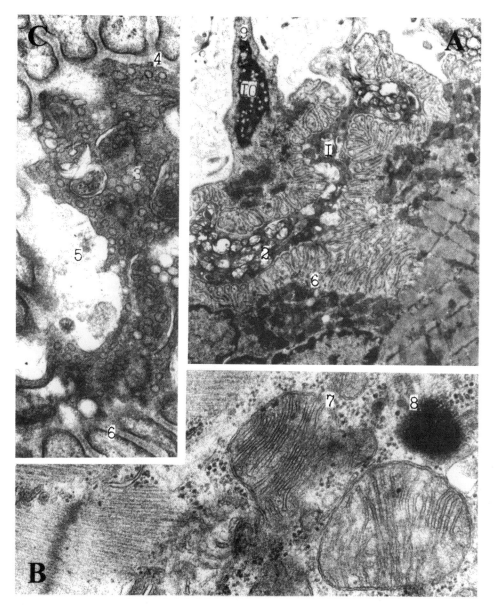

Figure 6. Electron micrographs of two neuromuscular junctions (1) on the fast-oxidative-glycolytic muscle fiber. Mitochondria (2), synaptic vesicles (3), synaptic cleft (4), dilated synaptic cleft (5), postsynaptic folds (6), mitochondria of the post synaptic zone (7), lysosome (8), muscle fiber nuclei (9), nucleus pores (10). Magnification, A-x 5000, B-x 52000, C-x 42000.

less clearly manifested. We were not able to observe essential structural changes in the nerve muscle synapses of the muscle fibers. Both the control and myopathic muscles possess glycolytic fibers that have characteristic elongated terminals and a faintly manifested postsynaptic region. In the axon terminals there occurs a quantity of smooth-surfaced synaptic vesicles and coated vesicles and mitochondria. Similar to the fast- oxidative-glycolytic muscle fibers there also occur in the glycolytic fibers many-layered concentric formations. On some occasions splitting of these layered membranes into vesicles can be observed. It is possible that this is one of the ways synaptic vesicles are formed (Figure 7).

Part of the terminals contain neurofilaments and lots of postsynaptic folds which branch off. The postsynaptic space is narrow and contains few mitochondria, some glycogen granules, secondary lysosymes and muscle cell nuclei the surface of which is activated. These nuclei form protruding and porous structural nuclei (Figure 8).

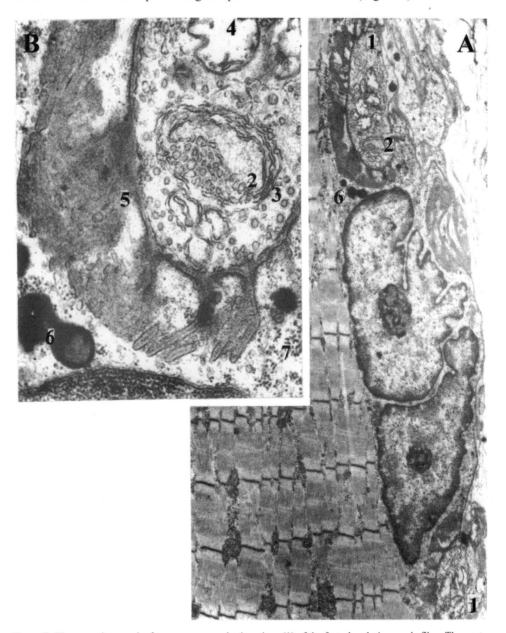

Figure 7. Electron micrograph of two neuromuscular junctions (1) of the fast-glycolytic muscle fiber. The postsynaptic region is narrow. In the axonal terminal vesiculation of membrane structures (2), coated synaptic vesicles (3), and mitochondria (4), postsynaptic folds (5), lysosome-like bodies (6) and glycogen granules can be seen (7). Magnification, A-x 9000, B-x 50000.

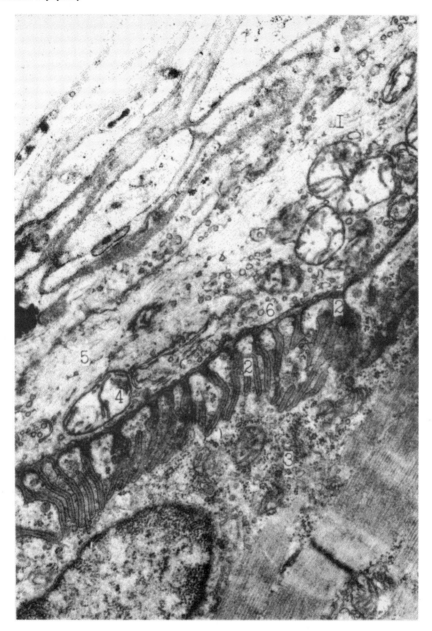

Figure 8. Electron micrograph of a fragment of neuromuscular junction (1) of fast-glycolytic muscle fiber. Axonal terminal is relatively long with several mitochondria (4), neurofilaments (5) and few synaptic vesicles (6). Postsynaptic folds (2) are regular postsynaptic sarcoplasm (3) contains a few organelles. Magnification, x 35000.

5. SIMILARITIES BETWEEN EXERCISE AND GLUCOCORTICOID MYOPATHIES

It is well known that administration of large doses of glucocorticoids induces muscle myopathy. The ultrastructural study showed the disarray of thick myofilaments in fast-twitch glycolythic fibers in dexamethasone-treated rats. Obviously the destructive process of myofilaments begins from the periphery of myofibrils, spreads to the central part of sarcomere near the H-zone and is distributed over all the A-band (9).

We get the impression that the thin filament and Z-line are much more resistant to any catabolic action of dexamethasone than thick filaments in glycolythic muscle fibers. The myofibrils of glycolythic muscle fibers are thinner in dexamethasone-treated rats. This is caused by the splitting of myofibrils. In oxidative-glycolythic fibers the destruction of thick myofilaments was remarkably less pronounced. Myofibrils were structurally normal in the dexamethasone-treated *soleus* muscle. There was evidence of increased lysosomal activity in the glycolythic muscle fibers and in the satellite cells of dexamethasone-treated rats. This manifested itself in increased numbers of lysosomes, autophagic vacuoles, which contain mitochondria and other membranous structures. In glycolythic fibers from the corticosteroid-treated rats the sarcoplasmic reticulum was frequently more prominent than in the controls. There are only slight morphological changes in oxidative-glycolythic fibers and there are no structural changes in slow-twitch muscle fibers during hormone treatment.

5.1. Catabolic Action of Glucocorticoids

The main reason for the development of muscle atrophy in corticosteroid myopathy is the accelerated catabolism of muscle proteins. It is well established that lysosomal and non-lysosomal pathways also exist in skeletal muscle to account for the degradation of their intracellular proteins. As the content of lysosomes in skeletal muscle is relatively low, the non-lysosomal pathway makes a particularly significant contribution and may be of special importance in the initial rate-limiting steps in the catabolism of myofibrillar proteins and consequently in the regulation of their turnover rate. A promoted breakdown of contractile proteins myosin and actin in the muscle during glucocorticoid administration was shown by the enhanced excretion of 3-methylhistidine (13). The slow-twitch skeletal muscles are much more resistant than the fast-twitch muscles to any catabolic action of the corticosteroids (8).

5.2. Role of Alkaline Proteinase in Intracellular Catabolism

The catabolic effect of glucocorticoids on myofibrils seems to be realised through the augmented alkaline proteinase activity (9). On the other hand, it is possible that the myosin heavy chains and actin are more sensitive to the action of alkaline myofibrillar protease than are the myosin light chains, at least under *in vitro* conditions (12). As has been pointed out, the weight reduction in different muscle types in dexamethasone-treated animals is in full accordance with the augmentation of their alkaline protease activity (8). Although a significant weight reduction occurred only in the fast-glycolytic fibers, the intracellular catabolic effect of glucocorticoids was noted in the fast oxidative-glycolytic and slow oxidative muscle fibers as well.

Alkaline proteases are synthesized in mast cells. After degranulation the enzyme enters the muscle cell, but the mechanism is unknown. Upon administration of large doses of glycocorticoids there is an increase in the number of mast cells in the perivascular porous connective tissue of the muscle fibers. Around the fast glycolytic muscle fibers, degranulation of mast cells is very clearly expressed. Simultaneously the number of mast cells in the lymph node medulla is considerably decreased. The lymph nodes are probably the sources of the muscle mast cells. This may imply that the increased number of mast cells may be the result of their migration from the lymph nodes. Forthy eight hours after glucocorticoid administration the mast cell number in lymph nodes is returned to control levels. In the atrophying muscle, myofibrillar destruction starts from those myosin filaments which are located in the peripheral part of the myofibrills (9). Thick filaments separate from the adjacent ones, bend and are obviously lysed. The actin filaments and Z-line seem to be more resistant to the action of alkaline protease, in comparison with the myosin filament (9). Muscle weakness in the case of glucocorticoid myopathy is most probably caused by lesions of the myofibrillar apparatus in the muscle fibers and by changes in the state of the neuromuscular synapses (9).

5.3. Prevention of Intracellular Catabolism by Exercise

The catabolic action of glucocorticoids on skeletal muscle was found to depend on the functional activity of skeletal muscles (8). Several studies have demonstrated that endurance exercise with simultaneous glucocorticoid treatment is an effective measure in retarding skeletal muscle atrophy (3, 4) and provides protection against one of the major effects of glucocorticoids, i.e. muscle wasting (1). Endurance exercise with simultaneous glucocorticoid treatment in rats causes a decrease of muscle mass, but in comparison with unexercising animals the mass of extensor *digitorum longus* and *gastrocnemius* muscles was subsequently 57 and 48 % higher.

In exercising glucocorticoid-treated rats the synthesis rate of actin in fast oxidative-glycolytic muscle was 65 % and in fast glycolytic fibers 49 % more intensive than in hormone-treated animals. At the same time synthesis of myosin heavy chain in fast glycoclytic fibers was 14 % more intensive. It seems that exercise causes an anabolic effect in both types of fast-twitch muscle fibers mainly by intensification of actin synthesis rate. In exercising glucocorticoid-treated animals significant intensification in the synthesis rate of myosin heavy chain was only observed in fast glycolytic fibers. Glucocorticoids can inhibit or diminish the action of androgens, and androgenic action may partly be involved in limiting or inhibiting glucocorticoid effects (3).

The effect of physical activity in inducing a less pronounced catabolic action of corticosteroids seems to be caused by the elevation of anticatabolic activity of exercise (8). Also, the role of endogenous androgens cannot be excluded from the anticatabolic effect of the exercise, since only moderate exercise had an anticatabolic activity in skeletal muscles of corticosteroid myopathic rats. Exhaustive exercise, on the contrary, augmented the catabolism in skeletal muscles (8). At the same time, anabolic steroid in these situations elevated the anticatabolic activity in the fast-twitch glycolytic muscle fibers. Fast-twitch glycolytic fibers are as sensitive to the catabolic action of glucocorticoids as to the anabolic action of anabolic steroids (9).

6. SUMMARY

It seems that there is some similarity between the structural changes in fast-glycolytic and fast-oxidative-glycolytic muscle fibers in case of exercise myopathy, although destructive changes in the glycolytic muscle fibers seem to be of smaller scope. On the one hand, this can be explained by the fact that glycolytic fibers do not participate so actively in the low-intensive exercise. On the other hand, the fairly large range of destructive processes in these fibers shows that glycolytic muscle fibers participate in long-term low-intensive exercise performing the junction of the skeletal muscle in the role of an organ.

It is well known, that fast-twitch muscle fibers are more sensitive to the action of corticosteroids. Therefore it is probable that the increase of endogenous corticosteron level in endurance-type exhausted rats and cortisol level in overtrained athletes may be the important factor in the pathogeneses of the exercise myopathy.

REFERENCES

1. Czerwinski S, Kurowski T, O'Neill T, Hickson R (1987) Initiating regular exercise protects against muscle atrophy from glucocorticoids. J Appl Physiol 63: 1504–1510
2. Foster C, Lehmann M (1997) Overtraining syndrome. In: Guten GN (ed) Running injuries. Saunders Philadelphia: 173–188
3. Hickson R, Davis J (1981) Partial prevention of glucocorticoid induced muscle atrophy by endurance training. Am J Physiol 241: E226-E232
4. Hickson R, Kurowski T, Andrew G, Cappacio J, Chatterton R (1986) Glucocorticoid cytosol binding in exercise-induced sparing of muscle atrophy. J Appl Physiol 60: 1413–1419
5. Järva J, Alev K, Seene T (1997) The effect of autografting on the myosin composition in skeletal muscle fibers. Muscle & Nerve 20: 718–727
6. Lehmann M, Foster C, Keul J (1993) Overtraining in endurance athletes: a brief review. Med Sci Sports Exercise 25: 854–862
7. Rusko H (1995) Recovery and overtraining. In: Viitasalo I and Kujala U (ed) The Way to Win Hakapaino Helsinki: 181–186
8. Seene T, Viru A (1982) The catabolic effect of glucocorticoids on different types of skeletal muscle fibers and its dependence upon muscle activity in interaction with anabolic steroids. J Steroid Biochem 16: 349–352
9. Seene T, Umnova M, Alev K, Pehme A (1988) effect of glucocorticoids on contractile apparatus of rat skeletal muscle. J Steroid Biochem 29: 313–317
10. Seene T, Alev K (1991) Effect of muscular activity on the turnover rate of actin and myosin heavy and light chains in different types of skeletal muscle. Int J Sports Med 12: 204–207
11. Seene T, Umnova M (1992) Relations between the changes in the turnover rate of contractile proteins, activation of satellite cells and ultra-structural response of neuromuscular junctions in the fast-oxidative-glycolytic muscle fibers in endurance trained rats. Basic and Applied Myology 2: 34–46
12. Seene T (1994) Turnover of skeletal muscle contractile proteins in glucocorticoid myopathy. J Steroid Biochem Molec Biol 50: 1–4
13. Seene T, Umnova M, Alev K, Puhke R, Kaasik P, Järva J, Pehme A (1995) Effect of overtraining on skeletal muscle of the ultrastructural and molecular level. In: Viitasalo J and Kujala U (ed) The Way to Win Hakapaino Helsinki: 187–189

10

MONITORING OVERLOAD AND REGENERATION IN CYCLISTS

Uwe A. L. Gastmann and Manfred J. Lehmann

Medizinische Universitätsklinik und Poliklinik
Abt. Sport- und Leistungsmedizin
Steinhövelstr. 9, D-89075 Ulm, Germany

1. INTRODUCTION

Training of high-performance athletes is always like a "ride on a razor blade". A balance of necessary phases of overload and recovery will, in general and if all other conditions are all right, e. g. day form, environmental conditions, etc., lead to good competition results. A training/competition >> recovery imbalance, however, will result in performance incompetence and overtraining syndrome.

In the past years many scientific working groups around the world are engaged in various aspects of overtraining. For an overview we like to refer to the review articles of Fry et al. (9) and Lehmann et al. (21, 23). An overtraining state originating in a training/competition >> recovery imbalance may happen quite often (4, 7, 14, 15, 17, 19, 24, 26, 29, 30) since a common reaction of many athletes and coaches to poor performance in competition or training is an increase in training loads with inappropriate regeneration. As a result of a stress >> recovery imbalance, carbohydrate deficit (5, 6), autonomic imbalance (16, 17), neuroendocrine imbalance (1, 2, 10–13, 22, 31, 32, 34), catabolic > anabolic imbalance (1), and amino acid imbalance (25, 27, 28, 33) have been assumed. Non-training stressors, e. g. social, educational, occupational, economical, nutritional, travel (8, 21), and monotony of training (3, 8) can worsen this state with large individual variations.

In this sense, the returning to baseline levels of the above mentioned mechanisms after a certain time of recovery, whereas exact baseline levels must not be totally achieved, could indicate an appropriate regeneration of the organism. However, scientific results especially of the duration necessary for regeneration are still poor. In general, athletes and coaches determine regeneration periods subjectively according to their experience. Beside the duration of the regeneration cycle, the grade of physical activity during the recovery period as well as the characteristics of the preceding training/competition, i.e., volume, intensity, and duration, or the individual athlete must be taken in consideration.

Overload, Performance Incompetence, and Regeneration in Sport, edited by Lehmann *et al.*
Kluwer Academic / Plenum Publishers, New York, 1999.

2. OVERLOAD AND REGENERATION IN UNTRAINED ATHLETES

To elucidate some aspects of regeneration processes, we undertook three controlled and prospective studies with cyclists. In a preliminary study (22), 7 recreational athletes (age 26±1 years, height 185±9 cm, body mass 79±7 kg, VO2max 55±6 ml x kg^-1 x min^-1) performed a cycle ergometer training in the laboratory 6 days a week for 6 weeks followed by a training-free period (recovery) of 3 weeks. The total exercise duration was 40–60 min a day including warm-up and cool-down with 4 days a week steady state (30 min; 89–96% of 4 mmol lactate performance [4LT]) and 2 days a week interval training (3–5 x 3–5 min; 110–127% of 4LT), which was up to 6 times of their baseline training levels. This unaccustomed intensive and monotonous training in untrained athletes resulted primarily in an increase in submaximum (4LT: 242±47 vs 262±48 W, p<0.001) and maximum (333±73 vs 342±80 W) power output after 3 weeks. However, after 6 weeks of training, the athletes were completely demotivated, and the athletes' performance stagnated or deteriorated. This happened even after 3 weeks of regeneration.

2.1. The Hypothalamo-Pituitary-Adrenocortical Axis

To evaluate training-related alterations in the hypothalamo-pituitary-adrenocortical system, we performed CRH stimulation tests (100 μg human corticorelin) in the early afternoon, when ACTH and cortisol levels show rather a circadian low. Baseline ACTH levels before training were in the normal range with slightly elevated cortisol levels according to the results of Wittert et al. (34). After the unaccustomed and intensive training, pituitary ACTH release was ~60% higher but with a ~10% lower adrenal cortisol release as an indicator of a reduced sensitivity of the adrenal cortex to ACTH. Even after 3 weeks of recovery, cortisol release was further reduced (~30%) by similar ACTH levels. Barron and coworkers (2) obtained similar results in overtrained marathoners who also had an impaired function of the hypothalamo-pituitary-adrenal axis which, in contrast to our results, "normalized" again after 4 weeks of recovery.

3. OVERLOAD AND REGENERATION IN ELITE JUNIOR CYCLISTS

The second and third study was done with junior amateur cyclists (n=7, age 17±1 years, height 179±2 cm, body mass 64±5 kg, VO2max 56±7 ml × kg^-1 × min^-1; n=10, age 17±1 years, height 181±4 cm, body mass 68±3 kg) of a german state team after an 8 month training and competition season one year apart, respectively. After a hard and long-lasting training/competition season with a tight competition schedule, the cyclists were at least at borderline of an overtraining state. The second study ("cycling") was planned with a synchronization cycle of 2 weeks with further overload followed by 5 weeks of active regeneration with indvidually given plans of reduced training. However, because of motivational reasons and the lack of sufficient controls, the athletes reduced their cycling volume only to ~52% the first 2 weeks and to ~42% the following 3 weeks of the "regeneration cycle" compared to the preceding synchronization cycle (median 435 km/wk, 50%ROC 226–561) where training was already reduced in relation to "normal" volumes during the training/competition season. Therefore, during the 5 weeks of recovery in the third study ("non-cycling"), only unspecific physical exercise, e. g. sport lessons in school, leisure time swimming, etc., was permitted which ran up to 4:20–5:10 hours a week.

In the cycling group, the maximum power output (median 366.7 W, 50%ROC 350–375) obtained from a graded cycle ergometer test (100 W/50 W/3 min) fell for ~5% after 2 weeks of inappropriate training reduction and stagnated after 3 more weeks of slightly further training reduction. The submaximum performance (4LT) showed a similar behavior. In a 10-km outdoor time trial, the race time stagnated with an in tendency faster race with a lower lactate difference after 5 weeks of recovery (15:15/9.9 vs 15:23/11.2 vs 15:05 min/8.5 mmol × l^-1). The achieved average speed of below 40 km × h^-1 for the 10-km course must be seen as an indicator of an overtraining situation. In contrast, the non-cycling group showed a slightly increasing maximum power output (median 340 W, 50%ROC 320–347) after 2 and after 5 weeks (~2%) of regeneration. See Figure 1.

3.1. The Hypothalamo-Pituitary-Adrenocortical Axis

Baseline resting ACTH (median 4.8 pmol × l^-1, 50%ROC 2.9–6.0) and cortisol levels (median 257 mmol × l^-1, 50%ROC 229–441) of the cycling group were in the normal range with an appropriate release in the CRH stimulated pituitary-adrenal function test. After the first 2 weeks of only slight training reduction, cortisol release was only ~50% by over 70% CRH release on an in total elevated level of these hormones. Only when the training volume was further reduced for 3 more weeks, ACTH and cortisol resting levels and release returned to similar values as at baseline (see Figure 2).

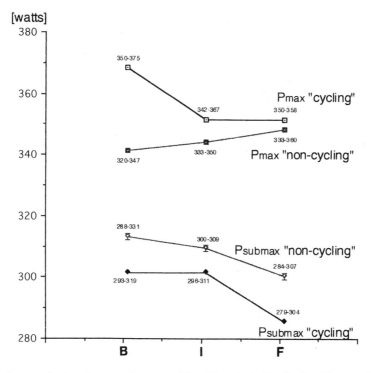

Figure 1. Maximum and submaximum performance of the athletes in the "cycling" and "non-cycling" regeneration study at baseline (B), interim (I), and final (F) examination. See text.

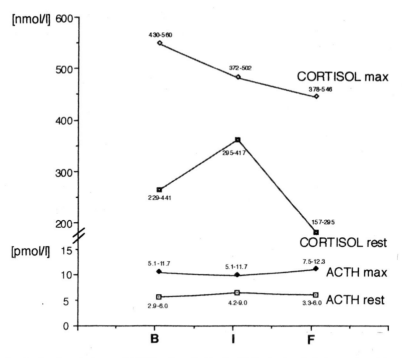

Figure 2. Resting levels and release of ACTH and cortisol in the CRH stimulated pituitary-adrenal function test at baseline (B), interim (I), and final (F) examination of the "cycling" study. See text.

3.2. The Autonomic Nervous System

As an other example of how the organism protects itself against overload dependent damage by impaird transmission of ergotropic signals to target organs, as we demonstrated above for the hypothalamo-pituitary-adrenal axis, we want to show some results of the autonomic nervous system. Whereas free plasma catecholamines represent more the actual stress-related sympathetic response, nocturnal urinary catecholamine excretion is seen as an indicator of the intrinsic sympathetic activity, because activating stimuli are widely excluded (19). Hooper and coworkers (14, 15) found elevated plasma noradrenaline levels in overtrained swimmers. In the "cycling" study, free plasma noradrenaline levels increased after 2 weeks of inappropriate training reduction and stagnated on this elevated level after 3 more weeks of only slightly further training reduction at the same submaximum (300 W) and maximum workloads (see Figure 3). The corresponding plasma glucose levels were about the same or reduced after 5 weeks of still to much training. Assuming similar energy stores, these results can be seen as an indicator of a reduced catecholamine sensitivity due to a reduced betareceptor density which was also observed in distance runners and swimmers during prolonged and intensive training (18).

After the syncronisation cycle in the second study, the cyclists had low nocturnal urinary noradrenaline concentrations which were in the range of overtrained elite cyclists (19), distance runners (20) or semiprofessional football players (19). The reduced training led to a "normalization" of the nocturnal catecholamine excretion already after 2 weeks and further more after the following 3 weeks (see Figure 4).

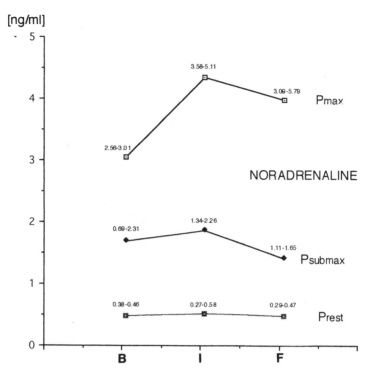

Figure 3. Plasma noradrenaline at rest, submaximum and maximum workload in a graded cycle ergometer test at baseline (B), interim (I), and final (F) examination of the "cycling" study. See text.

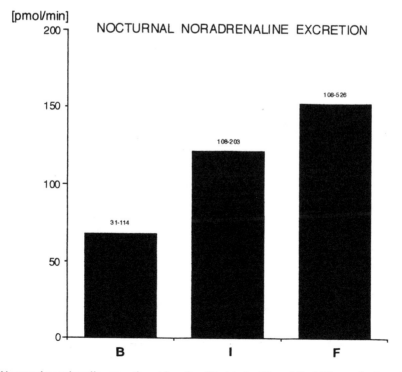

Figure 4. Nocturnal noradrenaline excretion at baseline (B), interim (I), and final (F) examination of the "cycling" study. See text.

4. CONCLUSION

In summary, some systems of the organism seem to react to overload with hyposensitivity of target organs as a protection against possible damage. However, as we could demonstrate above, the extend and the time course of different systems to the same overload exposure varies. The duration necessary for recovery after prolonged and intensive endurance training and/or competition is rather more than 2 weeks than under. During regeneration periods, training has to be markedly reduced to only low-intensity, low-volume units. However, exact scientific results are still lacking to supply athletes and coaches with practicable and reliable tools for training control.

REFERENCES

1. Adlercreutz H, Harkonen K, Kuoppasalmi K, Naveri H, Huthamieni H, Tikkanen H, Remes K, Dessipris A, Karvonen J (1986) Effect of trianing on plasma anabolic and catabolic steroid hormones and their response during physical exercise. Int J Sports Med 7 (Suppl) : 27–28
2. Barron JL, Noakes TD, Lewy W, Smith C, Millar RP (1985) Hypothalamic dysfunction in overtrained athletes. J Clin Endocrinol Metabol 60 : 803–806
3. Bruin D (1994) Adaptation and overtraining in horses subjected to increasing training loads. J Appl Physiol 76 : 1908–1913
4. Budgett R (1990) Overtraining syndrome. Br J Sports Med 24 : 231–236
5. Costill DL, Bowers R, Braunam G, Sparks K (1971) Muscle glycogen utilization during prolonged exercise on successive days. J Appl Physiol 31 : 834–838
6. Costill DL, Flynn MG, Kirwan JP, Houmard JA, Mitchell JB, Thomas R, Sung HP (1988) Effects of repeated days of intensified training on muscle glycogen and swimming performance. Med Sci Sports Exerc 20 : 249–254
7. Counsilman JE (1955) Fatigue and staleness. Athletic J 15 : 16–20
8. Foster C, Lehmann M (1996) Overtraining Syndrome. In: Guten GN (ed) Running Injuries. W. B. Saunders, Philadelphia-London-Toronto-Montreal- Sidney-Tokyo : 173–188
9. Fry RW, Morton AR, Keast D (1991) Overtraining in Athletes. An Update. Sports Med 12 (1) : 32–65
10. Hackney AC, Sinning WE, Bruor BC (1990) Hypothalamic-pituitary-testicular axis function in endurance-trained males. Int J Sports Med 11 : 298–303
11. Hackney AC (1991) Hormonal changes at rest in overtrained endurance athletes. Biology of Sport 8 : 49–56
12. Hakkinen K, Pakarinen A, Alen M, Kauhanen H, Komi PV (1987) Relationships between training volume, physical performance capacity and serum hormone concentrations during prolonged weight training in elite weight lifters. Int J Sports Med 8 (Suppl) : 61–65
13. Hakkinen K, Pakarinen A, Alen M, Kauhanen H, Komi PV (1988) Daily hormonal and neuromuscular responses to intensive strength training in 1 week. Int J Sports Med 9 : 422–428
14. Hooper SL, Mackinnon LT, Gordon RD, Bachmann AW (1993) Hormonal responses of elite swimmers to overtraining. Med Sci Sports Exerc 25 : 741- 747
15. Hooper SL, Mackinnon LT, Howard A, Gordon RD, Bachmann AW (1995) Markers for monitoring overtraining and recovery. Med Sci Sports Exerc 27 : 106–112
16. Israel S (1976) Zur Problematik des Übertrainings aus internistischer und leistungsphysiologischer Sicht. Medizin und Sport 16 : 1–12
17. Kuipers H, Keizer HA (1988) Overtraining in elite athletes. Sports Med 6 : 79- 92
18. Jost J, Weiss M, Weicker H (1989) Unterschiedliche Regulation des adrenergen Rezeptorsystems in verschiedenen Trainingsphasen von Schwimmern und Langstreckenläufern. in: Böning D, Braumann KM, Busse MW, Maassen N, Schmidt W (Hrsg) Sport - Rettung oder Risiko für die Gesundheit. Deutscher Ärzteverlag, Köln : 141–145
19. Lehmann M, Schnee W, Scheu R, Stockhausen W, Bachl N (1992) Decreased nocturnal catecholamine excretion: parameter for an overtraining syndrome in athletes? Int J Sports Med 13 : 236–242
20. Lehmann M, Gastmann U, Petersen KG, Bachl N, Seidel A, Khalaf AN, Fischer S, Keul J (1992) Training - overtraining: performance and hormone levels after a defined increase in training volume vs. intensity in experienced middle- and long-distance runners. Br J Sports Med 26 : 233–242

21. Lehmann M, Foster C, Keul J (1993) Overtraining in endurance athletes. A brief review. Med Sci Sports Exerc 25 (7) : 854–862
22. Lehmann M, Knizia K, Gastmann U, Petersen KG, Khalaf AN, Bauer S, Kerp L, Keul J (1993) Influence of 6-week, 6 days per week, training on pituitary function in recreational athletes. Br J Sports Med 27 : 186–192
23. Lehmann MJ, Lormes W, Opitz-Gress A, Steinacker JM, Netzer N, Foster C, Gastmann U (1997) Training and Overtraining: an overview and experimental results in endurance sports. J Sports Med Phys Fitness 37: 7 - 17
24. Morgan WP, Brown DR, Raglin JS, O'Connor PJ, Ellickson KA (1987) Psychological monitoring of overtraining and staleness. Br J Sports Med 21 : 107–114
25. Newsholme EA, Parry-Billings M, McAndrew N, Budgett R (1991) A biochemical mechanism to explain some characteristics of overtraining. In: Brouns F (ed) Advances in nutrition and top sport. Med Sport Sci 32, Karger, Basel : 79–83
26. Owen IR (1964) Staleness. Phys Educ 56 : 35
27. Parry-Billings M, Bloomstrand E, McAndrew N, Newsholme N, Newsholme EA (1980) A comunicational link between skeletal muscle, brain, and cells of the immune system. Int J Sports Med 11 (Suppl 2):122–128
28. Parry-Billings M, Budgett R, Koutedakis Y, Bloomstrand E, Brooks S, Williams C, Calder PC, Pilling S, Baigrie R, Newsholme EA (1992) Plasma amino acid concentration in the overtraining syndrome: possible effects on the immune system. Med Sci Sports Exerc 24 : 1353–1358
29. Stone MH, Keith RE, Kearney JT, Fleck SJ, Wilsond GD, Triplett NT (1991) Overtraining. A review of signs, symptoms and possible causes. J Appl Sport Science Research 5 : 35–50
30. Verma SK, Mahindroo SR, Kansal DK (1978) Effect of four weeks of hard physical training on certain physiological and morphological parameters of basketball players. J Sports Med 18 : 379–384
31. Vervoorn C, Quist AM, Vermulst LJM, Erich WBM, de Vries WR, Thijssen JHH (1991) The behavior of the plasma free testosterone/cortisol ratio during a season of elite rowing training. Int J Sports Med 12 : 257–263
32. Wheeler GD, Singh M, Pierce WD, Epling SF, Cumming DC (1991) Endurance training decreases serum testosterone levels without change in LH pulsatile release. J Clin Endocrin Metabol 72 : 422–425
33. Wilson W, Maughan RJ (1993) Evidence for the role of 5-hydroxytryptamine (5- HT) in fatigue during prolonged exercise. Int J Sports Med 14 : 297–298
34. Wittert GA, Livesey JH, Espiner EA, Donald RA (1996) Adaption of the hypothalamopituitary adrenal axis to chronic exercise stress in humans. Med Sci Sports Exerc 28 (8) : 1015–1019

11

MONITORING REGENERATION IN ELITE SWIMMERS

Sue L. Hooper and Laurel T. Mackinnon

Department of Human Movement Studies
The University of Queensland
Brisbane, Australia

1. INTRODUCTION

Regeneration from the negative aspects of intense training is an important focus of the taper period prior to major competition. For nearly half a century, competitive swimmers have been attempting to optimise competitive performance by training close to the point of overtraining and then regenerating while the training load is tapered. Tapering is a regeneration technique during which the physiological and psychological stresses of daily training are gradually reduced prior to competition in order to maximise the difference between the positive and negative effects of training.

The taper allows supercompensation processes to maximise the positive effects of training and regeneration processes to eliminate fatigue and other negative effects. For elite athletes to reach peak performance for a specific competition, it is desirable that coaches and sport scientists are able to accurately monitor regeneration occurring during the taper. Through careful monitoring procedures, peaking for elite competition can be deliberately programmed. However, as yet, reliable indicators for monitoring regeneration from intense training have not been identified.

The need for coaches to have available reliable monitoring tools for the taper period is similar to the need for monitoring tools for preventing overtraining. In each case, these tools allow the coach to titrate training loads so that performance is optimised. A number of studies have sought to identify variables which reliably identify and prevent overtraining syndrome.[9,17,18,31,34] For example, in our previous work, a wide range of parameters were measured at rest and after submaximal and maximal effort swimming at peak training loads for the season. A self-rating of stress kept daily in a training log and plasma catecholamine levels predicted the extent of overtraining (an overtraining score) ($r^2 = .85$) [9]. In the same study, the swimmers' improvement in competitive performance in National Trials (from previous best times) was predicted by measures taken at one time point dur-

Overload, Performance Incompetence, and Regeneration in Sport, edited by Lehmann *et al.*
Kluwer Academic / Plenum Publishers, New York, 1999.

ing the taper. The mean percent improvement for all race times was predicted by four ratings of well-being kept in a daily training log (i.e., measures of sleep, fatigue, stress, and muscle soreness) ($r^2 = .72$).

This work suggested that easily measured variables such as well-being and other physiological and psychological markers may provide objective means of monitoring regeneration during the taper. To date, there have been very few studies which have examined the technique in a realistic athletic environment or attempted to determine markers for successful tapering. For example, there is little empirical evidence on how the taper is best programmed considering that many different methods of reducing the workload are available. Training volume, frequency and intensity can all be altered and the duration of the taper extended from one to several weeks.[14] To more fully understand the factors contributing to successful tapering, studies focusing on the responses of athletes to different types and lengths of tapers are needed. Moreover, the identification of variables for monitoring regeneration during tapering would assist in tracking the positive changes which result from this process.

The aim of the following two studies was therefore to further investigate methods of monitoring swimmers' regeneration during tapering and to compare different types of tapering regimes.

2. OVERVIEW OF TWO STUDIES

Both studies assessed tapering as currently used by state and national level Australian swimmers. Swimmers were used since they have traditionally relied on tapering to allow regeneration after intense training and thus peaking for major competition. The first study attempted to identify easily measured variables which predicted the improvement in performance as a result of tapering. Measurements of performance, tethered swimming forces, mood states and self-ratings of well-being were obtained before and after tapering for State Titles. Data from this study, together with our other data[13] led us to conclude that accurate prediction of improved performance as a result of tapering is possible.

The second study assessed different types of tapers by comparing three regimes currently used by Australian competitive swimmers. The tapers included reductions in training frequency, volume, and volume and intensity combined. They were implemented in a realistic competitive situation as the regeneration phase before State Titles. The data suggested that the exact type of work load reduction may not be critical as long as sufficient opportunity to recover is provided.

3. STUDY ONE: NON-INVASIVE MONITORING OF REGENERATION

The findings of our previous studies (9,13) prompted us to undertake work with a larger group of swimmers to determine whether non-invasive measures could accurately predict performance improvement as a result of regeneration with tapering. This was considered possible since previous work had demonstrated that competitive performance improvement after tapering could be predicted from daily self-ratings of well-being taken three days before competition.[9] Further, biochemical measures such as plasma creatine kinase and serrum ferritin concentrations and erythrocyte and leukocyte counts have been consistently shown to be unreliable as markers of overreaching or overtraining.[9,19,28]

Changes in biochemical variables are likely to reflect an acute response to a single exercise session rather than a general reflection of the athlete's overtraining or recovery status. [5,9,32]

3.1. Methods

This study compared changes in various measures from before to after tapering with the change in performance time from before to after tapering. Parameters measured included 400m maximal effort swim time, mood states (POMS), well-being measures reported in daily training logs, and forces during tethered swimming. Eighteen swimmers ranked in the top 20 in the state volunteered for the study. There were seven males and 11 females in the group (mean age and SD = 16.3 ± 1.0 years). Swimmers were tested twice: first after 12 weeks of training and before tapering commenced, and again after two weeks of tapering for State Titles. Subjects were familiarised with all procedures in pre-study practice sessions and provided written informed consent forms. The study was approved by the University of Queensland's Ethics Committee and the subjects were free to withdraw at any time.

The taper was programmed by each swimmer's coach without interference from the researchers. For the week before and during the taper, swimmers kept a daily log of training details including distance swum, time spent on gym work, and training intensity. Training intensity and ratings of well-being were rated on a seven point scale as described in previous work[8,9] (e.g., 1 = very, very easy; 2 = very easy; 3 = easy; 4 = neither easy nor hard; 5 = hard; 6 = very hard; 7 = very, very hard for training intensity). Ratings of well-being including fatigue, stress, quality of sleep and muscle soreness were rated as 1 = very, very good; 2 = very good; 3 = good; 4 = neither good nor bad; 5 = bad; 6 = very bad; 7 = very, very bad.

Testing was conducted at the same time of day and at the same training venue for each swimmer. The Profile of Mood States (POMS)[21] was completed on arrival at the pool during seated rest. After a standardised warm up (800m easy swimming and stretching exercises as each swimmers wished), forces generated during maximal effort tethered swimming were measured as described previously.[10] After 400m easy swimming to recover and 10min seated rest, subjects swam 400m freestyle at maximal effort. This was timed by stop watch to 0.01s.

3.2. Results

Figure 1 shows the training details for the week before tapering commenced (42km swimming at an average intensity rating of 5.1 on the scale of 1 to 7 and 2.5h gym work) and for the second week of tapering (21km at an intensity of 4.7 with 0.8h gym work). Swimming distance and gym work, but not intensity, were significantly lower ($p<0.05$) during tapering. The change from before to after tapering was calculated for each variable as a percent change (Table 1). Log book measures recorded daily (training variables and well-being ratings) were averaged for the week prior to testing to give a single value for statistical analysis. The improvement in 400m swim time during tapering was regressed on all other variables using a step-down regression analysis.[33] Performance improvement (%) with tapering (Y) was predicted by the following percent changes:

$$Y = -1.2 + 0.3\text{depression} + 0.5\text{force} - 1.3\text{muscle soreness} - 0.1\text{anger} + 0.4\text{confusion} + 0.4\text{log book fatigue} - 0.1\text{vigour} \ (r^2 = .71)$$

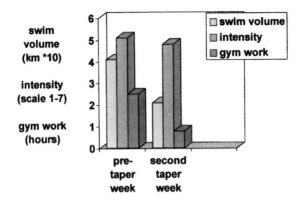

Figure 1. Training details for study one. Changes in swim volume and gym work significant (p<0.05).

Percent changes in swimming forces, POMS measures of depression, anger, confusion and vigour and the daily log book measures of fatigue and muscle soreness, together, were able to account for 71% of the variance of the percent performance time changes. Each of the performance changes predicted by this model was not significantly different from that predicted by a larger battery of variables and gave a prediction significantly better than the mean change for the 18 swimmers.

3.3. Discussion

The main finding of the study was the battery of variables which predicts regeneration during tapering for competition. This battery provides the coach with the possibility of easily establishing which athletes may not be recovering appropriately to peak for competition. All variables in the battery are non-invasive and require minimal analysis and therefore there is little delay between measurement and results. Furthermore, except for peak tethered swimming force, the variables require little equipment with only a daily log

Table 1. Percent changes (mean and SD) from before to after tapering in Study One

Measure	Change (%)
400m swim time	− 2.6 ± 7.9
Sleep quality	+0.5 ± 1.0
Fatigue	− 0.1 ± 1.1
Stress	− 0.2 ± 0.7
Muscle soreness	− 0.1 ± 0.7
POMS	
Tension	− 1.0 ± 5.1
Depression	− 2.7 ± 6.5
Anger	− 2.0 ± 5.6
Vigor	− 1.1 ± 8.0
Fatigue	− 3.1 ± 6.3
Confusion	− 0.6 ± 3.0
Total Mood Disturbance	−12.0 ± 28.8
Peak Force	− 2.1 ± 2.3

book and a questionnaire required, although the POMS must be administered by a registered psychologist. Therefore, these variables appear to be of potential assistance to the coach in programming the taper so that performance is optimised in competition.

Performance improvement in 400m time as a result of tapering (2.6 %) was similar to the 2.8 to 7.0% changes reported in previous research in runners, swimmers and cyclists.[2,3,14,15,16,20,24,30] The change in tethered swimming force was also similar to the improvements in muscular strength and power previously demonstrated with tapering. [6,16,20,30] While this improvement has not been shown to account for the improvement in performance on its own,[10] the inclusion of force as a monitoring tool for regeneration is consistent with previous work.[2,30] The inclusion of well-being measures is also consistent with our previous work which showed that improvement in competitive swimming performance from previous best times was predicted by four ratings of well-being recorded in a daily training log (i.e., measures of sleep, fatigue, stress, and muscle soreness).

Psychological measures were also included in the prediction battery. This was not surprising since mood states have been shown to be useful in monitoring training loads.[23,25,27,34] Furthermore, POMS measures have been shown to improve as a result of tapering after intense training.[4,11,27,35]

Together, peak swimming force, POMS measures of depression, anger, confusion and vigour and the well-being ratings of fatigue and muscle soreness predicted the improvement in swimming times with tapering. These variables accounted for 71% of the variance of the percent performance time changes. This was substantially lower than that demonstrated in our previous study where invasive measures were included.[13] In this previous study,[13] changes were measured in more than 40 physiological, biomechanical and psychological variables from before to after two weeks tapering for National Titles. Ten elite swimmers had measures taken of resting and post-exercise heart rate, blood pressure, and lactate concentration; resting hormonal levels of cortisol, free testosterone and catecholamines; self-assessment of well-being and mood states; and forces during tethered swimming. Regression analysis showed that plasma norepinephrine concentration, heart rate after maximal effort swimming (HRmax), and confusion as measured by the Profile of Mood States (POMS) predicted the improvement in swimming times with tapering ($Y = 14 + .0051$norepinephrine $+ .0790$HRmax $- .0258$confusion; $r^2 = .98$).

The results of this previous study, together with our current data, indicate that accurate prediction of performance with tapering is possible in swimmers. However, further work is needed to determine whether it is possible to extend these findings to other situations and other types of athletes. At present, it appears that monitoring of regeneration prior to competition may be performed using a range of easily measured variables including well-being and mood states.

4. STUDY TWO: COMPARISON OF DIFFERENT TYPES OF TAPER REGIMES

This study focused on the success of different types of workload reduction during the taper. Research data on this aspect of tapering are limited. To our knowledge, only two studies[15,30] appear to have compared taper regimes but these were quite different from those used by competitive swimmers. For example, competitive swimmers do not use tapers which require either complete rest or a change in exercise mode as implemented in the above two studies. The purpose of this study was to compare the benefits of three tapering techniques similar to those currently used by Australian swimmers peaking for important competition: (i) reduction in training frequency; (ii) reduction in training volume; and (iii) reduction in training volume and intensity combined.

4.1. Methods

Twenty-seven swimmers training and tapering for State Titles (mean age and SD = 15.0 ± 1.6 years) were chosen for the study. After a minimum of 12 weeks training with their usual coaches, they commenced four weeks of training according to a set program which included 40km swimming and 150min dryland work completed in 8 workouts per week. Coaches and assistants on the pool deck used the scale of 1 to 7 (described above) to maintain the intensity of training as close as possible to a rating of 5 for all swimmers. At the end of the four weeks the swimmers were divided into three groups with equal numbers of males and females in each group. Each group (n=9) tapered according to one of three taper regimes. At the end of the taper period the subjects competed in State Championships. Tapering lasted two weeks (10 training days and 4 rest days).

The reduced frequency taper was individualised for each swimmer according to ratings of well-being recorded in the daily log (as described above). Titrating training according to psychometric factors has been reported previously.[1,22,29] The workouts were the same as those of the last week of intense training but in first taper week the afternoon workout was deleted if any of the early morning ratings were >5 (on the 1–7 scale) and in second taper week the afternoon workout was deleted if any ratings were >4. The second type of taper also used the training program of the last week of intense training but the distance swum in each section of the program was reduced progressively by 10% on each day for 10 days. The third taper reduced volume as in the second taper and also progressively replaced 10% of the intense training with easy swimming so that on the last taper day all the swimming was at low effort.

The swimmers were tested four times: before and after the 4 weeks standardised, intense training and after one and two weeks of tapering. Performance in 100m and 400m freestyle (swim time), tethered swimming forces, mood states and self-ratings of well-being were measured as in Study One. Statistical comparisons were made using one way analysis of variance with time as a repeated measure.

4.2. Results

Training details for the three taper groups are shown in Figure 2. For the 27 swimmers, neither one nor two weeks of tapering resulted in significant improvements in the times for the 100m and 400m swims. However, significant improvements ($p<0.05$) were observed in the POMS measures of tension, depression and anger after one week of tapering and in total mood disturbance (TMD), fatigue, and peak tethered swimming force after two weeks (Figures 3 and 4). Non-significant improvements were demonstrated in the daily subjective ratings of quality of sleep, stress, fatigue and muscle soreness. There were no significant differences for any variable among the three tapering techniques.

4.3. Discussion

To our knowledge, this is the first study to compare different taper regimes similar to those practised by elite competitive swimmers and culminating in important competition. The tapers did not result in a statistically significant improvement in swim time. However, this lack of significant improvement has previously been reported in other studies on reduced training load or tapering[30,35] and our previous work with this type of athlete

swim volume (km x 10), intensity (scale 1-7), gym work (hours)

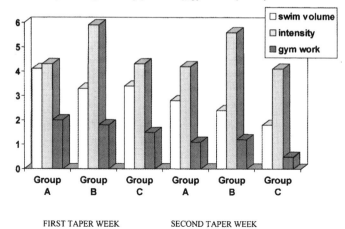

FIRST TAPER WEEK   SECOND TAPER WEEK

Figure 2. Training details for study two: Group A = reduced training frequency, Group B = reduced training volume, Group C = reduced training volume and intensity; no significant changes (p>0.05).

has repeatedly shown improvements of less than 2% as a result of tapering.[9,12] The differences are likely to be related to the different pre-taper training loads and the different types, durations and workloads of the tapers employed in the various studies.

Significant improvements in POMS measures have previously been demonstrated with tapering[4,11,27] and an improvement in mood states is considered a beneficial result of tapering since successful athletes demonstrate fewer mood disturbances than unsuccessful athletes.[7,23,26] It may be expected that some mood states (e.g., tension) may not decline with tapering since the reduction in workload with tapering is coincident with an increase in psychological stress as competition approaches.[27] However, in this study significant improvements were observed in all POMS variables except vigor and confusion. It is unclear why these two variables did not improve significantly as did the other POMS variables. The maintenance of high training intensity in the three tapers may have been a factor since POMS variables have been shown to be more closely related to training intensity than training volume.[11] Figure 3 clearly shows that there was little decrease in intensity in the tapers and this reflects current practice by competitive swimmers. The swimmers in this study may therefore have needed longer than the two weeks allowed to recover fully for competition.

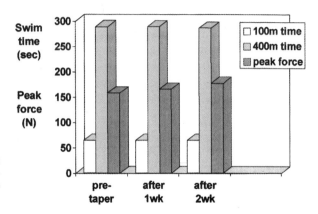

Figure 3. Time and force changes with tapering: peak force after two weeks tapering significantly greater than pre-taper (p<0.05).

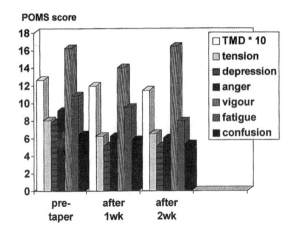

Figure 4. Changes in POMS meaures with tapering. TMD and fatigue after two weeks tapering are significantly less than pre-taper (p<0.05). Tension, depression, and anger after one and two weeks tapering are significantly less than pre-taper (p<0.05).

The results also showed that one week of reduced training was not long enough to maximise the benefits of tapering, since two weeks showed greater improvements in several variables (e.g., total mood disturbance, fatigue and swimming forces) compared with one week. The lack of improvement in the subjective ratings of well-being in the daily logs also supports the suggestion that the tapers did not allow sufficient opportunity for full recovery. Significant improvements in measures of well-being have previously been demonstrated with tapering.[9] However, this previous study monitored recovery in elite swimmers after very high training loads (e.g., mean weekly swim volumes ranging from 46.0 to 62.3km) which resulted in some swimmers becoming overtrained. A significant improvement in well-being was therefore expected in this previous study as a result of an increased capacity to recover and a longer taper (mean = 2.4wks) compared with the present study (pre-taper training 38km and taper duration 2wk). In the present study, the 27 swimmers showed a significant decline in swim volume and gym work with tapering but intensity was maintained at the pre-taper level. Maintaining the intensity level at the pre-taper level may result in swimmers requiring two weeks rather than one week for full recovery for competition.

None of the three types of tapers were shown to be more beneficial than the others. This may not be surprising considering that the training for the three regimes was very similar (see Figure 2). Furthermore, the subject numbers in each group were small and some variables had small effect sizes (e.g., performance time). In summary, tapering resulted in significant improvements in tethered swimming forces and self-reported mood states, which can be considered important benefits of the process, but may not be reflected in statistically significant performance improvements in training or competition. Different methods of reducing the workload during tapering may be equally effective in achieving the benefits of tapering if sufficient time is allowed for regeneration to occur. For some swimmers, more than two weeks of tapering appears to be necessary for complete recovery before competition.

5. CONCLUSION

Tapering provides an important opportunity for athletes to regenerate after intense training. It allows the negative effects of training to dissipate. Measures of well-being such as fatigue and muscle soreness and measures of performance (eg., time and forces) show significant improvements as long as sufficient time is allowed for recovery. It appears that at least two weeks is needed for competitive swimmers. The method used to reduce the workload may not be critical, but maintenance of high training intensity may require longer taper durations (>2wk).

Monitoring the swimmer in order to predict the success of the taper is an important requirement for accurately titrating the training load so that the swimmer performs optimally in competition. The first study showed that a few easily measured variables such as tethered swimming forces, mood states and well-being ratings kept in daily log books could provide a feasible monitoring procedure for swimmers. The second study demonstrated the need for such monitoring since one week of work load reduction was insufficient to allow complete regeneration for the swimmers. Further work appears warranted to determine whether these findings can be extended to other types of athletes training and competing in different environments with different performance demands.

REFERENCES

1. Berglund B, Safstrom H (1994) Psychological monitoring and modulation of training loads of world-class canoeists. Med Sci Sports Exerc 26:1036–40
2. Costill DL, King DS, Thomas R, Hargreaves M (1985) Effects of reduced training on muscular power in swimmers. Physician Sportsmed 13:94–101
3. Costill DL, Thomas R, Ropbergs RA, Pascoe A, Lambert C, Barr S, Fink WT (1991) Adaptations to swimming training: influence of training volume. Med Sci Sports Exerc 23:371–377
4. Cogan KD, Highlen PS, Petrie TA, Sherman WM, Simonsen J (1991) Psychological and physiological effects of controlled intensive training and diet on collegiate rowers. Int J Sports Psych 22:165–180
5. Flynn MG, Pizza FX, Boone JB, Andres FF, Michaud FA, Rodriguez-Zayas JR (1994) Indices of training stress during competitive running and swimming seasons. Int J Sports Med 15:21–26
6. Gibala MJ, MacDougall JD, Sale DG (1994) The effects of tapering on strength performance in trained athletes. Int J Sports Med 15:492–497
7. Guttmann MC, Pollock ML, Foster C, Schmidt D (1984) Training stress in Olympic speed skaters: A psychological perspective. Physician Sportsmed 12:45–57
8. Hooper SL, Mackinnon LT, Gordon RD, Bachmann AW (1993) Hormonal responses of elite swimmers to overtraining. Med Sci Sports Exerc 25:741–747
9. Hooper SL, Mackinnon LT, Howard A, Gordon RD, Bachmann AW (1995) Markers for monitoring overtraining and recovery. Med Sci Sports Exerc 27:106–112
10. Hooper SL, Mackinnon LT, Wilson BD (1995) Biomechanical responses of elite swimmers to staleness and recovery. Aust J Sci Med Sport 27:10–14
11. Hooper SL, Mackinnon LT, Hanrahan S (1997) Mood states as an indication of staleness and recovery. Int J Sport Psych 28:1–12
12. Hooper SL, Mackinnon LT, Ginn EM (1998) Effects of three taper techniques on the performance, forces, and psychometric measures of competitive swimmers. Eur J Applied Physiol 78:258–263
13. Hooper SL, Mackinnon LT, Howard, A Norepinephrine as a marker of recovery after intense training. (submitted)
14. Houmard JA (1991) Impact of reduced training on performance in endurance athletes. Sports Med 12:380–393
15. Houmard JA, Scott BK, Justice CL, Chenier TC (1994) The effects of taper on performance in distance runners. Med Sci Sports Exerc 26:624–631
16. Johns RA, Houmard JA, Kobe RW, Hortobagyi T, Bruno NJ, Wells IM, Shinebarger, MH (1992) Effects of taper on swim power, stroke distance and performance. Med Sci Sports Exerc 24:1141–1146
17. Lehmann M, Deickhuth HH, Gendrisch G, Lazar W, Thum M, Kaminski R, Aramendi JF, Peterke E, Wieland W, Keul J (1991) Taining-overtraining: A prospective experimental study with experienced middle- and long-distance runners. Int j Sport Med 12:444–452
18. Lehmann M, Baumgartl P, Wiesenack C et al. (1992) Training-overtraining: influence of a defined increase in training volume vs training intensity on performance, catecholamines and some metabolic parameters in experienced middle- and long-distance runners. Eur J Appl Physiol 64:169–177

19. Mackinnon LT, Hooper SL, Jones S, Gordon RD, Bachmann AW (1997) Hormonal, immunological and hematological responses to intensified training in elite swimmers. Med Sci Sports Exerc 29:1637–1645
20. Martin DT, Scifres JC, Zimmerman SD, Wilkinson JG (1994) Effects of interval training and a taper on cycling performance and isokinetic leg strength. Int J Sports Med 15:485–491
21. McNair DM, Lorr M, Droppleman LF (1971) EDITS Manual for the Profile of Mood States. Educational and Industrial Testing Services, San Diego, pp 1–29
22. Morgan WP (1994) Psychological components of effort sense. Med Sci Sports Exerc 26:1071–1077
23. Morgan WP, Brown DR, Raglin JS, O'Connor PJ, Ellickson KA (1987) Psychological monitoring of overtraining and staleness. British J Sports Med 21:107–114
24. Mujika I, Busso T, Lacoste L, Barale F, Geyssany A, Chatard J-C (1996) Modeled responses to training and taper in competitive swimmers. Med Sci Sports Exerc 28:251–258
25. O'Connor PJ, Morgan WP, Raglin JS (1991) Psychobiologic effects of 3d of increased training in female and male swimmers. Med Sci Sport Exerc 23:1055–1061
26. Prapavessis H, Berger B, Grove JR (1992) The relationship of training and pre-competitive mood states to swimming performance: an exploratory investigation. Aust J Sci Med Sport 24:12–17
27. Raglin JS, Morgan WP, O'Connor PJ (1991) Changes in mood states during training in female and male college swimmers. Int J Sports Med 12:585–589
28. Rowbottom DG, Keast D, Goodman C, Morton AR (1995) The haematological, biochemical and immunological profile of athletes suffering from the overtraining syndrome. Eur J Appl Physiol 70:502–509
29. Rushall BS (1990) A tool for measuring stress tolerance in elite athletes. Appl Sport Psych 2:51–66
30. Shepley B, MacDougall JD, Cipriano N, Sutton JR, Tarnopolsky MA, Coates G (1992) Physiological effects of tapering in highly trained athletes. J Appl Physiol 72:706–711
31. Snyder AC, Kuipers H, Cheng B, Servais R, Fransen E (1995) Overtraining following intensified training with normal muscle glycogen. Med Sci Sports Exerc 27:1063–1070
32. Urhausen A, Gabriel H, Kindermann W (1995) Blood hormones as markers of training stress and overtraining. Sports Med 20:251–276
33. Veldman DJ (1967) Fortran Programming for the Behavioural Sciences. New York: Rinehart and Winston pp 281–295
34. Verde T, Thomas S, Shephard RJ (1992) Potential markers of heavy training in highly trained distance runners. British J Sports Med 26:167–175
35. Wittig AF, Houmard JA, Costill DL (1989) Psychological effects during reduced training in distance runners. Int J Sports Med 10:97–100

12

OVERLOAD AND REGENERATION DURING RESISTANCE EXERCISE

Andrew C. Fry [*]

Human Performance Laboratories
The University of Memphis
Memphis, Tennessee 38152

1. OVERLOAD VS. OVERTRAINING

Whether they know it or not, anyone who has entered a weight training facility with the intention of getting bigger or stronger or improving physical performance has had to deal with the concept of an overload. This includes not only athletes, but also recreational lifters, those exercising for health reasons, and anyone going through a rehabilitation program. In order to alter the physiological system, it is necessary to expose the system to an unaccustomed stress. Only in this manner can the body respond by appropriately adapting to the stress. As with many stresses, the adaptation response is not immediate, and requires a period of regeneration. How long this regeneration period must be to permit complete recovery will depend on the training stress itself. In many ways, this training stress response is quite similar to the General Adaptation Syndrome as proposed by Hans Selye.[33] The recovery time required following a training stress is directly related to the magnitude of the stress. In training terms, this stress is called an overload, or exposing the body to a stress beyond which it is accustomed.[2]

All effective training programs, whether resistance exercise or some other training modality, include a certain level of overload. The challenge to those prescribing the exercise program is to properly titrate the overload. Too much of an overload stress and the ability to recover is compromised. Extreme overload may even result in the phenomenon called overtraining where the ability to perform work decreases.[14,30,36,41] While impaired performances due to overtraining can occur with any type of exercise, it should be noted that overtraining with resistance exercise may not be identical to overtraining due to other types of exercise.[4,6,24,38,42] In addition, a lesser form of overtraining called overreaching can

_____

[*] Address correspondence to: Andrew C. Fry, Ph.D., 135 Roane Field House, The University of Memphis, Memphis, Tennessee 38152. Phone: 901-678-3479; Fax: 901-678-3464; E-mail: fry.andrew@coe.memphis.edu

be recovered from with a short recovery period of several days.[4,6,14,30,36] If carefully prescribed, overreaching can be an effective training method, provided the coach or trainer knows when to properly remove such an overload. Regardless of the degree of the stress, whether it is the stress from a properly designed exercise program, or it is overtraining, all exercise stresses designed to enhance performance provide a certain level of overload.

2. TYPES OF COMPETITIVE RESISTANCE EXERCISE

When concerned with the stress response to resistance exercise, it is critical to note how different the various forms of resistance exercise can be. One place to start is with the various forms of competitive resistance exercise sports. Specifically, these include Olympic-style weightlifting, power lifting, and body building.[2,26,39] Success in each of these sports depends on the proper use of resistance exercise, but each sport has its own unique characteristics. Olympic-style weightlifting is characterized by large muscle mass exercises that generate extremely high levels of power.[15,26] The competitive lifts include the snatch lift and the clean & jerk lift. Power lifting also incorporates large muscle mass exercises, but the emphasis is on maximal force production, while power production is considerably lower than for weightlifting.[15,21,26] The competitive lifts include the squat, the bench press, and the dead lift. Since near maximal forces are generated, the velocity of movement during the power lifts is necessarily slow, in contrast to the high velocities of the weightlifting movements. Body building is concerned primarily with muscle hypertrophy which should be presented in an esthetically pleasing manner.[40] Training for body building competition utilizes both large muscle mass exercises as well as small muscle mass exercises to create the proper balance in muscular size. Training for each of these sports involves extensive use of heavy resistance exercises, but training for each sport is quite different, and the results are extremely different. The types of training stresses experienced for each sport result in a unique overload environment for each type of training. As a result, although all three of these sports can be categorized as competitive lifting sports, the overload and regeneration requirements are distinctly different.

By far, the majority of individuals who regularly perform resistance exercise are not competitive lifters. Many people use resistance exercise to assist training for other sports, for general fitness and health, for recreational enjoyment, or for clinical rehabilitation purposes.[2,39] Each of these training purposes requires different types of programs to attain the desired results. As a result, it is important to realize that not all resistance exercise programs are identical, and that each program presents its own unique physiological demands and overload qualities. To further complicate the problem, athletes using resistance exercise to supplement training for their particular sport must concern themselves with additional sources of overload. Not only do they experience the overload from the resistance exercise, but also the stresses from their sport-specific training.[24,28] For many sports, this includes training for both the aerobic and anaerobic energy systems. Such a complex scenario can be difficult for a coach or athlete to deal with, since the resulting overload can be quite different for the various training modalities.

3. ACUTE RESISTANCE EXERCISE TRAINING VARIABLES

To fully appreciate how variable heavy resistance exercise prescriptions can be, it is necessary to closely study the components of a single exercise session. Only when this is done can the complex problem of long-term training prescription be properly addressed.[2,37,39] The components of a single training session are completely described by the five acute training variables;[2]

- Choice of exercise,
- Order of exercise,
- Load,
- Volume of exercise and
- Rest (i.e., between sets).

Whatever the specifics of the training session entail, it can always be described by these variables. Although this may seem to be a fairly simple list, close examination of each of the variables illustrates what a complicated situation this is.

3.1. Choice of Exercise

In the world of resistance exercise, there are literally thousands of exercises that can be performed. This includes not only the huge selection of resistance exercise machines that are on the market, but the even larger number of exercises that can be performed with free weights of various types. Another consideration includes the type of resistance employed, whether a dynamic constant external resistance (i.e., isotonic), isokinetic, isometric, hydraulic or pneumatic, elastic bands, etc. The type of muscle action can also have a profound impact on the training effect, be it concentric, eccentric, isometric, or any combination of these types of muscle actions. Furthermore, some exercises focus on the transition from one type of action to another (e.g., the amortization phase during plyometric exercises). Other considerations include the primary movers activated, the synergistic muscles required, and the biomechanical characteristics of the exercise. As illustrated by the force-velocity relationship, the velocity of movement also is a critical factor in determining the choice of exercise. Undoubtedly, numerous other factors are important considerations for choosing particular exercises, but it is readily apparent that there are easily thousands of potential exercises to chose from.[2,15,21,39,40]

3.2. Order of Exercise

Once the actual exercises have been selected, exercise sequence must be determined. Factors to consider include the following;

- Large muscle mass exercises before or after small muscle mass exercises,
- Multi-joint exercises before or after single joint exercises,
- Core exercises before or after assistance exercises,
- Free weight exercises before or after machine exercises,
- High skill exercises before or after low skill exercises, and
- Agonist-antagonist combinations or one muscle or muscle group at a time.

Just as with determining choice of exercise, there are numerous other considerations for determining the order of performing the exercises. Due to the biomechanical and physiological characteristics of the exercises, the sequence in which they are performed can greatly influence the resulting adaptation. For example, lets assume that barbell squats and knee extensions are to be used. One involves a large muscle mass while the other involves a small muscle mass, one involves multiple joints while the other uses a single joint, one is a core exercise while the other is often considered an assistance exercise. Depending on which exercise is performed first, acute fatigue from the first exercise will affect the metabolic environment, the neuromuscular recruitment patterns, and the load which can be correctly handled for the following exercise. Thus, the overload characteristics can vary simply due to the order of exercise.

3.3. Load

One of the first questions asked by those initiating a resistance exercise program is "How heavy a resistance should I use?" The training load is often defined as the exercise intensity. Just as aerobic exercise intensity is often prescribed as a percentage of maximal heart rate, resistance exercise is often prescribed as a percentage of maximal strength levels (i.e., % 1 repetition maximum [RM]). When defined in this manner, it is called relative intensity, since the load is relative to each individual's strength level.[2,39] On the other hand, absolute intensity is determined by the actual weight being moved.[39] For the purposes of resistance training records, the mean intensity is sometimes calculated, which is simply the average weight used for a particular series of exercises or an entire training session.[39] In many situations, relative intensity is used to determine the training load for an individual. Obviously, a 1 RM strength test must first be performed to insure an accurate loading, and this may not always be practical or possible. In addition, there is considerable inter-individual and inter-exercise variability in the number of repetitions that can be performed at any particular relative intensity.[22] When a coach/trainer is very familiar with their athlete/client, it is possible to accurately prescribe using relative intensity. In situations, however, where the coach/trainer is unable to accurately identify the proper relative intensity, RM loading is useful[2]. This is where the load is defined as the most weight an individual can use for a prescribed number of repetitions (e.g., 1 RM, 5 RM, 10 RM, etc.). Using this method, it is possible to prescribe an exact number of repetitions. The trade-off is that it may not be desirable to exercise at such a maximal level of effort on every set of every exercise for every training session. Each method of prescribing training load, however, can be very effective for producing the desired gains.

3.4. Volume of Exercise

Training volume for resistance exercise can be defined in several ways. In its simplest sense, training volume is simply the total number of repetitions performed during a training session.[2,39] When defined this way, it is important to remember that this calculation indicates nothing about quality of the training session, but simply indicates the quantity. Volume can also be quantified by determining the volume-load which is the number of repetitions x the weight on the bar or machine.[39] Volume-load can be a very helpful tool for tracking overload for an individual over a period of time. Perhaps an even more exact measure of the training volume performed by an individual is to calculate the total work in Joules. This includes not only the number of repetitions and the weight being moved, but also how far the weight was moved.[27,39] While this calculation may be critical in the research setting, it is generally more than is needed by the coach or athlete in a practical setting.

3.5. Rest Intervals

This important training variable is often overlooked when resistance training programs are designed. Among competitive athletes, it is well known that changing this one variable alone will have a tremendous impact on the resulting training stress. A quick re-

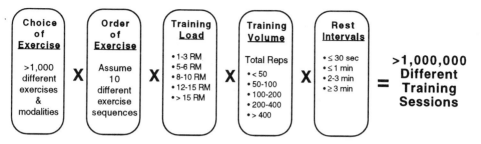

Figure 1. Manipulation of the five acute training variables can easily result in over one million individual resistance exercise training sessions.

view of typical training practices of competitive lifters reveals that inter-set rest intervals can vary from <30 sec to ≥3 min depending on the desired affect.[2,15,39,40] Such differences will not only impact the metabolic environment, but will significantly alter the acute hormonal profile.[27]

With all the choices available, it becomes readily apparent that there are numerous resistance exercise training protocols possible for any one session. The question is, how many are there, and what impact will each have on the desired adaptations? Figure 1 illustrates how the five acute training variables can be manipulated to result in over one million possible combinations. While it is likely that the physiological stresses of many of these possible training sessions will be quite similar, each one will be unique in one or more characteristics.

4. TRAINING PERIODIZATION

Thus far we have only considered how to vary a single resistance exercise training session. Long-term exercise programming, however, involves planning the training through various periods that may focus on particular goals of the training. Such planned training variation is called periodization, and has been shown to be particularly effective for eliciting positive adaptations.[2,36,37,39] Besides optimizing training adaptations, one of the goals of periodized training is the avoidance of overtraining. By properly planning the training ahead of time, the coach/trainer infuses periods of lesser training stress to allow adequate recovery following high training stress periods. All of the five acute training variables are carefully manipulated in this process, as well as several additional variables. For example, training frequency can be adjusted to result in greater or lesser training volumes over a specified period of time. When looking at the total resistance exercise program, not just one session, it is easy to see that the possible variety is even much greater than what is illustrated in Figure 1.

5. PHYSIOLOGICAL AND PERFORMANCE ASPECTS OF PERIODIZED TRAINING

At this point one might ask "Exactly how physiologically different are these various training regimens?" From a hormonal perspective, it has been shown that variations of both short-term[5,6,8,9,11,12,17,19,38] and long-term[6,9,16,24,36,38] training programs can result in markedly different hormonal profiles, indicative of the different physiological stresses.

Furthermore, when the five acute training variables are altered, the endocrine responses to single training sessions can be amazingly different.[20,27] Such changes in the hormonal environment have been shown to be correlated with changes in the fiber types and sizes in skeletal muscle due to resistance exercise.[35]

Training periods of excessive high intensity resistance exercise have resulted in suggestions of altered b-adrenergic sensitivity,[10,11] similar to what has been observed with endurance athletes.[31,32] Short periods of stressful resistance exercise have also resulted in altered neural regulation of muscle activity.[17] Although sometimes considered a long-term adaptation, skeletal muscle fiber types and areas, and protein expression can become apparent after as few as four training sessions,[34,35] indicative of how plastic skeletal muscle can be to a resistance exercise training stimulus.

Perhaps of most importance to the coach and athlete are the effects of different resistance exercise programs on physical performance. As has been previously mentioned, athletes in the various resistance exercise sports all train quite differently, resulting in the specificity needed for successful performance in their respective sports.[2,15,21,39,40,42] Simple changes in the resistance exercise prescription can readily change the focus of the training from that of maximal strength, to that of maximal power or maximal hypertrophy. In addition, many sports require various combinations of muscular strength, power, and size for optimal performance, thus requiring a carefully designed resistance exercise prescription to insure the proper results.

6. COMPARING RESISTANCE EXERCISE TO OTHER TYPES OF EXERCISE

A considerable amount of literature is available on the physiology of overtraining with endurance athletes.[14,30,31,41] As is typical of much research in the broad area of exercise science, most early research on this topic has dealt with aerobic activities and endurance athletes. Initial attempts to address the problems of overtraining in resistance exercise had to rely extensively on data generated from endurance athletes.[36] It was pointed out by van Borselen et al.[42] that since the physiological adaptations to appropriate resistance exercise were quite different from the adaptations with endurance exercise, likewise, the maladaptations accompanying resistance exercise overtraining were also likely to be different than those accompanying endurance overtraining. It thus became important to not only study the physiology of excessive overload during resistance exercise, but from a practical standpoint, to determine the exercise prescription variables contributing to this problem. Although much research to date has attempted to titrate out the contributing roles of either resistance exercise volume[36,38] or intensity,[4,6] it must be noted that these two variables can never be completely isolated. Proper resistance exercise prescription must always concern itself with the proper proportions of volume and intensity. In many real life settings, the problem with overtraining is actually a combination of too much training volume *and* too great a training intensity.

So, you may ask, what does all this information on resistance exercise program design have to do with overload and regeneration in resistance exercise? Before a complete understanding of regeneration and recovery is possible, one must be thoroughly familiar with the training stresses themselves. Just as Figure 1 shows how variable the training stresses can be, the manners in which the body responds to these stresses are also quite variable. When addressing the problem of overtraining or overreaching, it must be remembered that the characteristics of the problem are not always identical, and can be manifested in many different manners. The rest of this chapter will closely examine the scientific literature on overload and regeneration for resistance exercise, and will point out several practical applications of this information.

7. OVERLOAD DURING WEIGHTLIFTING

Although all types of resistance exercise are readily quantified, much of the scientific data on overload with resistance exercise has been derived from the sport of Olympic-style weightlifting. This is due primarily to the fact that, as an Olympic sport, weightlifting is generally more organized than the other competitive sports, and can be studied by monitoring national teams and training camps at national Olympic training centers. In addition, since weightlifting relies more on a structured coaching staff than do the other resistance exercise sports, such research can be coordinated with the coaches responsible for the training programs.

It has been observed with endurance exercise that the endogenous ratio of two steroid hormones, testosterone and cortisol, appear to be related to changes in the training stresses.[1] It was noted that as the stress of training increased, the testosterone/cortisol ratio (Tes/Cort) decreased. These two hormones were thought to represent the anabolic/catabolic environment of the body. From an endocrinologist's view, this is an oversimplification of a very complex hormonal system, and the Tes/Cort ratio is most likely indicative of numerous other physiological factors as well. What is interesting to note, however, is that this ratio has repeatedly been associated with alterations in the training environment, and has been suggested as a potential indicator of overload or overtraining.[1,6,8,9,14,16,17,18,19,24,-30,31,36,38,41] On the other hand, it should be noted that changes in the Tes/Cort ratio do not always accompany either endurance overtraining[31] or high intensity resistance exercise overtraining.[4,6] As such, it is not appropriate to use this simple ratio alone to identify an overtrained state.

Long-term weightlifting training of elite weightlifters over two years has been shown to increase resting levels of testosterone, thus influencing Tes/Cort.[18] Such augmentation of testosterone occurred despite the fact that these athletes were already highly trained and were already competitive at the national and international levels. When these athletes were studied over a six week period, it was noted that resting testosterone concentrations were negatively related, and resting cortisol concentrations were positively related to the volume-load (reps x weight). The net result was that when volume-load increased, Tes/Cort decreased, and when volume-load decreased, Tes/Cort was restored.[16] The decrease in the volume-load represented the training taper immediately prior to a major weightlifting competition, thus, an enhanced Tes/Cort ratio may be desirable for optimal strength and power performances. When even shorter periods of training were monitored, one week of increased training volume (250 - 500% increases) resulted in decreases in the Tes/Cort ratio.[8] Overall, it appears that with increases in weightlifting training volume, the Tes/Cort ratio responds in a manner similar to endurance exercise when training volume is increased.[14,30,31,41] An interesting finding with one week of increased volume weightlifting is that oral supplementation with branched-chain amino acids (i.e., leucine, isoleucine, valine) had no effect on weightlifting performance or the Tes/Cort ratio as an indicator of training stress.[8] These findings fail to support the central fatigue hypothesis suggested as a contributing factor for overtraining.[29] This topic will be addressed in greater detail elsewhere in this text.

In a sport like weightlifting, almost all the sport-specific training occurs in the weightlifting hall. But what about sports that include resistance exercise as only a supplemental part of the total training program? This actually includes most sports, where strength training is performed as only one part of the total program, and both aerobic and anaerobic capacities must be optimized. The problem with overtraining for these sports may depend on finding the proper balance between anaerobic and aerobic training, since it is easy to end up with too great a total training volume. Recent data indicates that training programs that include both high intensity endurance and resistance exercise may increase the risk of overtraining as indicated by the Tes/Cort ratio.[28] Coaches and athletes must be aware that training programs from sports that perform only resistance exercise may not be appropriate where additional types of exercise are included.

8. REGENERATION FACTORS DURING WEIGHTLIFTING

Although excessive overload for resistance exercise can be induced by increasing training volume and/or intensity, there appear to be several strategies available to counter-act some of these problems.

8.1. Training Program

The most obvious strategy, of course, is a properly designed training program.[2,15,37,38] This includes not only the individual training sessions, but also the long-term training program. Periodization techniques allow for training periods of high stress as well as periods for recovery. Such recovery is a planned phase of the training, and is scheduled before the athlete reaches a state of overtraining. It should also be noted that recovery does not necessarily mean cessation from training, but is usually characterized by decreases in training volume and/or intensity. In modern athletic society, many coaches and athletes are tempted to look for the "secret ingredient" for avoiding overtraining, whether it is a nutritional supplement, a different training device, or some other "new gimmick". The bottom line is that the most important factor is undoubtedly a properly designed training program.

8.2. Assessment Program

At a recent coaching symposium on overtraining for the top U.S. Olympic Team coaches, it became readily apparent that most of the coaches routinely utilized simple field tests to determine when their athletes were being pushed too hard. Such assessments required careful record keeping by the coach for each athlete involved. The tests have to be simple to administer, and have to be applicable to the specific sport. Such an approach is easier to use for sports that are highly quantifiable, such as swimming, cycling, weightlifting, and track & field. In sports that are not as readily quantifiable, the importance of close communication between the coach and athlete is essential. Examples of these sports include wrestling, gymnastics, and numerous team sports. Nothing can replace the subjective assessment of a conscientious coach who is in close contact with each athlete they work with. Even in sports where large numbers of athletes are included on a single squad (e.g., American football), it is possible for coaches to develop an objective and/or subjective method for determining the training status of their athletes.

8.3. Prior Exposure to Stressful Training

Study of the U.S. National Junior Weightlifting Squad over a several years period has resulted in much insight into the phenomenon of overload and overtraining in that sport.[5,6,8,9,25,38] With the cooperation of the coaching staff, a standardized weightlifting exercise session was administered to participants in a one week training camp.[25] In addition to physical performance measures, pre- and post-exercise serum samples were assessed for testosterone and cortisol, as well as several other endocrine variables. In agreement with previous resistance exercise literature,[27] the acute hormonal responses included elevations in both testosterone and cortisol, as well as decreases in the Tes/Cort ratio. Based on the previous findings of Häkkinen et al.,[18] athletes with ≥ 2 years weightlifting training experience were compared with those with < 2 years experience. Such comparisons revealed that the acute testosterone response of the more experienced weightlifters was significantly greater than for the less experienced weightlifters. Further comparison of these two groups showed no differences in physical maturity or weightlifting strength. These results suggest that the anabolic hormone response to a single weightlifting session may be enhanced with increased training experience.

In a related study, junior age-group weightlifters participated in a 4 week training camp which included increased training volume during the first week. Ten of these athletes had participated in a similar training camp the previous year,[8] and were returning after one year of additional training.[9] As expected, the 10 returning athletes significantly increased 1 RM lifting performances compared to the previous year. When their hormonal responses to the 1 week of increased training volume was compared to those athletes participating for the first time, it was discovered that the Tes/Cort ratio did not decrease, but was actually maintained. Combined with the fact that weightlifting performances were not adversely affected, this suggests that prior exposure to high volume training stresses increases one's tolerance for such stressful training. The take home message for coaches and athletes is that carefully monitored phases of very stressful training can increase an athlete's future ability to train at maximal levels, thus perhaps permitting them to more closely attain their genetic potential.

When the weightlifting performance and hormonal responses of the entire 4 week training camp were compared, further support of the important role of training experience and prior stressful training is evident.[5] A subset of more experienced weightlifters that had qualified for a senior level competition (i.e., the U.S. Olympic Festival) were compared to the rest of the athletes. Changes in the testosterone and cortisol responses during the 1 week of high volume training and the following 3 weeks of normal training were correlated with changes in competitive weightlifting performance. The less experienced weightlifters exhibited negative correlations between Tes/Cort and weightlifting performance during the first week ($r = -.70$) and positive correlations during the 3 weeks of normal training ($r = .51$). On the other hand, the more experienced weightlifters exhibited no correlation between Tes/Cort and weightlifting performance during the first week ($r = .00$) and highly positive correlations during the 3 weeks of normal training ($r = .92$). Such results again suggest that the more experienced weightlifters more readily tolerated stressful training, and in this case, responded more favorably to the decrease in training volume.

9. ROLE OF EXCESSIVE INTENSITY OVERLOAD DURING RESISTANCE EXERCISE

Up to this point, all the data presented has dealt with increases in resistance exercise training volume, but that is only one side of the coin. In some scenarios, too great a relative intensity is used for too long a duration.[4,6,10,11,12,42] To study overtraining resulting from excessive relative intensity, our laboratory has attempted to design a model of overtraining that would allow us to monitor the physiological and performance progression as a state of overtraining develops. Early attempts included training 5 days·week^{-1} using 8 × 1 at 90–95% of 1 RM on a squat-simulating machine.[7] Testing was performed on the 6th day each week, while the 7th day was used for rest. The net result was that 1 RM strength actually continued to improve, and overtraining did not occur. In retrospect, two factors may have contributed to the subjects' ability to tolerate this type of training. Lehmann et al.[31] have suggested that at least one day a week be used for recovery for endurance training, and the day of rest from the high intensity resistance exercise may have permitted enough recovery to permit continued gains in 1 RM strength. Another possible factor is that the relative intensity (90–95% 1 RM), although high, was low enough for this one exercise protocol to allow the subjects to tolerate the training. If so, subtle changes in relative intensity may be very critical when providing variation in a resistance exercise program.

Further attempts to induce overtraining due to excessive relative intensity were successful in attenuating 1 RM performances by approximately 11%, with recovery taking up to 8 weeks.[10] Resistance training was performed every day using 10 × 1 at 100% 1 RM on the squat-simulating machine. As previously suggested, changes in the relative intensity of 5–10% are critical when training in this range. The removal of the recovery day appeared to also increase the occurrence of overtraining. A surprising result of this study was that decreases in testosterone and the Tes/Cort ratio did not accompany the decreases in 1 RM strength.[12] Additionally, no changes were observed for free testosterone, % bound testosterone, the free Tes/Cort ratio, luteinizing hormone, adrenocorticotropic hormone, growth hormone, or creatine kinase. Each of these variables has been previously implicated as possibly accompanying overtraining.[2,14,30,36,41] Although the hormonal data may not agree with much of the the the data from high volume resistance exercise training[5,8,9,38] or endurance overtraining,[2,14,30,36,41] there was evidence of maladaptation in the peripheral muscle tissue[10] similar to what has been reported for endurance overtraining.[31,32] At the same time that exercise-induced epinephrine and norepinephrine increased[11], involuntary stimulated muscle force decreased.[10] As suggested by Lehmann et al. for endurance overtraining,[32] this may have been indicative of a decreased b$_2$ adrenergic sensitivity of the involved muscles.[11] Future study of this type of overtraining should focus on possible maladaptations of the peripheral muscles, as well as continued study of the autonomic nervous system. As has been suggested by Israel,[23] overtraining may produce an autonomic imbalance between the sympathetic and parasympathetic systems, and the augmented catecholamine concentrations observed during high intensity resistance exercise overtraining may be indicative of what has been termed sympathetic overtraining.

An attempt to modify the model of overtraining to a free weight environment has not been completely successful to date.[13] Barbell squats were performed for 5 × 1 at 90–95% 1 RM for 3 days·week^{-1} for 3 weeks. The result was a plateau in 1 RM strength despite the use of high relative intensities generally ideal for increasing 1 RM strength. Even though a decreased training frequency and lower relative intensity were used when compared to previous similar research,[7] 1 RM actually stagnated rather than continued to increase. These results demonstrate how different free weight resistance exercise is from machine resistance exercise, and suggest that training capacity with free weights may be less than for machine modalities.

An often overlooked factor of many resistance training programs is that of training monotony.[3,36] Many coaches and athletes end up using training programs that do not incorporate adequate variety. As was presented in the first part of this chapter, there is no excuse for this problem with resistance exercise since there are so many options. It is likely that a contributing factor to performance decrements in the high intensity overtraining studies was a lack of variety.[10,11,12] But how can this be measured? Foster and Lehmann suggest calculating the ratio of mean training load for a specified training period, and dividing this by the standard deviation of the training load.[3,31] This can be easily done for resistance exercise by calculating this from the volume-load. A high ratio indicates inadequate variation, while a low ratio indicates greater variation. Continual monitoring of this variable may provide a simple method to quantify monotonous training and help the coach and athlete avoid this problem.

10. PRACTICAL IMPLICATIONS

To date, the data available on resistance exercise overtraining suggest the following considerations:

- The endocrine environment appears to reflect both short- and long-term resistance exercise training stresses.
- High volume resistance exercise overload may present a hormonal environment similar to that observed with endurance overtraining.
- Prior exposure to high volume training periods and greater training experience enhances training tolerance and capacity.
- When training with maximal or near-maximal resistances, slight decreases in the relative intensity can greatly enhance recovery.
- Providing 1 or more recovery days per week can greatly enhance recovery.
- Avoiding training monotony is critical, and can be easily avoided by carefully monitoring training records.
- Carefully designed periodized training programs may be essential to avoid overtraining.

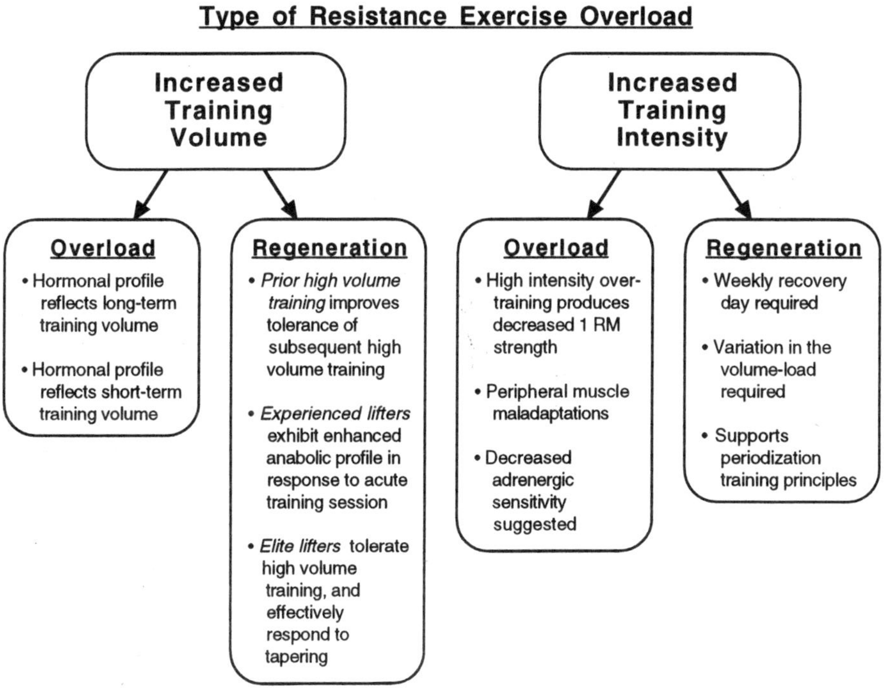

Type of Resistance Exercise Overload

Increased Training Volume		Increased Training Intensity	
Overload	**Regeneration**	**Overload**	**Regeneration**
• Hormonal profile reflects long-term training volume	• *Prior high volume training* improves tolerance of subsequent high volume training	• High intensity over-training produces decreased 1 RM strength	• Weekly recovery day required
• Hormonal profile reflects short-term training volume	• *Experienced lifters* exhibit enhanced anabolic profile in response to acute training session	• Peripheral muscle maladaptations	• Variation in the volume-load required
	• *Elite lifters* tolerate high volume training, and effectively respond to tapering	• Decreased adrenergic sensitivity suggested	• Supports periodization training principles

Figure 2. Types of resistance exercise overload and factors contributing to regeneration.

- Total work capacity may be less for free weights due to the different biomechanical and physiological nature of such exercise.
- High intensity resistance exercise overtraining may not be accompanied by the classic hormonal indicators of overtraining such as decreased testosterone and Tes/Cort, and elevated cortisol.
- Peripheral maladaptations of the involved skeletal muscle other than muscle damage may contribute to decreased force producing capabilities.

Figure 2 summarizes several important factors contributing to overload and regeneration during resistance exercise. Future research efforts should be designed to further study the overtraining implications of various training programs and modalities, as well as to determine the contributing physiological mechanisms in both the autonomic nervous system, as well as the peripheral musculature.

REFERENCES

1. Adlercreutz H, Härkönen M, Kuoppasalmi K et al. (1986) Effect of training on plasma anabolic and catabolic steroid hormones and their response during physical exercise. Int J Sports Med 7:S27-S28.
2. Fleck SJ, Kraemer WJ. (1997) Designing Resistance Exercise Programs, 2nd edition. Human Kinetics, Champaign, IL.
3. Foster C, Lehmann M. (1997) Overtraining syndrome. In: G.N. Guten (ed) Running Injuries. Saunders, Philadelphia: 173–188.
4. Fry AC. (1998) The role of training intensity in resistance exercise overtraining and overreaching. In: Kreider RB, Fry AC, O'Toole ML (eds) Overtraining in Sport. Human Kinetics, Champaign, IL: 107–127.
5. Fry AC, Koziris LP, Kraemer WJ et al. (in review) Hormonal and competitive performance relationships during an overreaching training stimulus in elite junior weightlifters. J Str Cond Res.
6. Fry AC, Kraemer WJ. (1997) Resistance exercise overtraining and overreaching - neuroendocrine responses. Sports Med 23(2):106–129.
7. Fry AC, Kraemer WJ, Lynch JM et al. (1994) Does short-term near-maximal intensity machine resistance exercise induce overtraining? J Appl Sport Sci Res 8:188–191.
8. Fry AC, Kraemer WJ, Stone MH et al. (1993) Endocrine and performance responses to high volume training and amino acid supplementation in elite junior weightlifters. Int J Sports Nutri 3:306–322.
9. Fry AC, Kraemer WJ, Stone MH et al. (1994) Endocrine responses to overreaching before and after 1 year of weightlifting. Can J Appl Physiol 19:400–410.
10. Fry AC, Kraemer WJ, van Borselen F et al. (1994) Performance decrements with high-intensity resistance exercise overtraining. Med Sci Sports Exerc 26:1165–1173.
11. Fry AC, Kraemer WJ, van Borselen F et al. (1994) Catecholamine responses to short-term high-intensity resistance exercise overtraining. J Appl Physiol 77:941–946.
12. Fry AC, Kraemer WJ, Ramsey LT. (in press) Pituitary-adrenal- gonadal responses to high intensity resistance exercise overtraining. J Appl Physiol.
13. Fry AC, Webber JL, Weiss LW, Li Y. (in press) Impaired performances with high intensity free weight resistance exercise. J Str Cond Res.
14. Fry RW, Morton AR, Keast D. (1991) Overtraining in athletes: an update. Sports Med 12:32–65.
15. Garhammer J, Takano B. (1992) Training for weightlifting. In: Komi PV (ed) Strength and Power in Sport. Blackwell Scientific, London: 357–369.
16. Häkkinen K, Pakarinen A, Alén M et al. (1987) Relationships between training volume,physical performance capacity, and serum hormone concentrations during prolonged training in elite weight lifters. Int J Sports Med 8:61–65.
17. Häkkinen K, Pakarinen A, Alén M et al. (1988) Daily hormonal and neuromuscular response to intensive strength training in one week. Int J Sports Med 9:422–428.
18. Häkkinen K, Pakarinen A, M. Alén et al. (1988) Neuromuscular and hormonal adaptations in athletes to strength training in two years. J Appl Physiol 65(6):2406–2412.
19. Häkkinen K, Pakarinen A. (1991) Serum hormones in male strength athletes during intensive short-term training. Eur J Appl Physiol 63:194–199.
20. Häkkinen K, Pakarinen A. (1993) Acute hormonal responses to two different fatiguing heavy-resistance protocols in male athletes. J Appl Physiol 74(2):882–887.

21. Hatfield FC. (1981) Power Lifting - A Scientific Approach. Contemporary Books, Chicago, IL.

22. Hoeger W. (1987) Relationship between repetitions and selected percentages of one repetition maximum. J Appl Sport Sci Res 1(1):11–13.

23. Israel S. (1996) Zur problematik des übertrainings aus internischer und leistungphysiologischer sicht. Medizin Sport 16:1–12.

24. Kraemer WJ. (1997) Factors involved with overtraining for strength and power. In: Kreider RB, Fry AC, O'Toole ML (eds) Overtraining in Sport, Human Kinetics, Champaign, IL: 69–86.

25. Kraemer WJ, Fry AC, Warren BJ, et al. (1992) Acute hormonal responses in elite junior weightlifters. Int J Sports Med 13:103–109.

26. Kraemer WJ, Koziris LP. (1994) Olympic weightlifting and power lifting. In: Lamb DR, Knuttgen HG, R Murray (eds) Physiology and Nutrition for Competitive Sports, Cooper Publishing, Carmel, IN: 16–54.

27. Kraemer WJ, Marchitelli L, Gordon SE et al. (1990) Hormonal and growth factor responses to heavy resistance exercise protocols. J Appl Physiol 69(4):1442–1450.

28. Kraemer WJ, Patton JF, Gordon SE et al. (1995) Compatibility of high-intensity strength and endurance training on hormonal and skeletal muscle adaptations. J Appl Physiol 78(3):976–989.

29. Kreider RB. (1998) Central fatigue hypothesis and overtraining. In: Kreider RB, Fry AC, O'Toole ML (eds) Overtraining in Sport, Human Kinetics, Champaign, IL: 309–331.

30. Kuipers H, Keizer HA. (1988) Overtraining in elite athletes: review and directions for the future. Sports Med 6:79–92.

31. Lehmann M, Foster C, Netzer N et al. Physiological responses to short- and long-term overtraining in endurance athletes. In: Kreider RB, Fry AC, O'Toole ML (eds) Overtraining in Sport, Human Kinetics, Champaign, IL: 19–46.

32. Lehmann M, Keul J, Wybitul K et al. (1982) Effect of selective and non-selective adrenoceptor blockade during physical work on metabolism and sympatho-adrenergic system. Drug Res 32:261–266.

33. Selye H. (1956) The Stress of Life. McGraw Hill, New York.

34. Staron RS. (1997) Human skeletal muscle fiber types: delineation, development, and distribution. Can J Appl Physiol 22(4):307–327.

35. Staron RS, Karapondo DL, Kraemer WJ et al. (1994) Skeletal muscle adaptations during early phase of heavy-resistance training in men and women. J Appl Physiol 76:1247–1255.

36. Stone MH, Keith RE, Kearney JT et al. (1991) Overtraining: a review of the signs and symptoms and possible causes. J Appl Sports Sci Res 5:35–50.

37. Stone MH, O'Bryant H, Garhammer J. (1981) A hypothetical model for strength training. J Sports Med Phys Fit 21:342–351.

38. Stone MH, Fry AC. (1998) Increased training volume in strength/power athletes. In: Kreider RB, Fry AC, O'Toole ML (eds) Overtraining in Sport, Human Kinetics, Champaign, IL: 87–105.

39. Stone MH, O'Bryant H. (1987) Weight Training - A Scientific Approach. Bellwether Press, Minneapolis.

40. Tesch PA. (1992) Training for body building. In: Komi PV (ed) Strength and Power in Sport. Blackwell Scientific, London: 370–380.

41. Urhausen A, Gabriel H, Kindermann W. (1995) Blood hormones as markers of training stress and overtraining. Sports Med 20:251–276.

42. van Borselen F, Vos NH, Fry AC et al. (1992) The role of anaerobic exercise in overtraining. Natl Str Cond Assoc J 14:74–79.

13

REGENERATION AFTER ULTRA-ENDURANCE EXERCISE

Mike I. Lambert, Alan St. Clair Gibson, Wayne Derman, and
Timothy D. Noakes

MRC/UCT Bioenergetics of Exercise Research Unit
Sport Science Institute of South Africa
P.O. Box 115
Newlands, 7700
South Africa

1. INTRODUCTION

Most of the research on endurance exercise has focussed on the physiological and metabolic demands during an event and on factors causing fatigue (7, 29, 33, 39). Fatigue during and after prolonged submaximal running may have many origins. The conventional explanation for fatigue after running a marathon or ultra-marathon is that glycogen is depleted in the skeletal muscles, thus reducing their ability to produce force (9, 38). This form of fatigue may occur only when the muscle glycogen concentrations fall below a critical level (3), perhaps causing impaired sarcoplasmic reticulum function (16). Fatigue, particularly after prolonged exercise exceeding 4 hours, may also coincide with the development of hypoglycaemia. There is some evidence to suggest that this fatigue may be delayed if carbohydrate is ingested during the event (36). The fatigue associated with hypoglycaemia probably has its origins in the central nervous system (CNS). Another mechanism underlying CNS fatigue after prolonged exercise proposed by Newsholme et al. (1992) is that as the duration of exercise increases, more plasma free-tryptophan crosses the blood brain barrier and is converted into serotonin, which increases the perception of fatigue (28). Fatigue after prolonged exercise, particularly in hot, humid environments may also be caused by factors affecting the CNS arising from sustained elevation of muscle and possibly brain temperature (32). Studies have also identified a type of neuromuscular fatigue which occurs during and after a marathon and which accounts for the decrement in running performance (30, 31). The above discussion highlights some of the metabolic issues and aetiology of fatigue associated with prolonged submaximal exercise. In contrast to the growing understanding of the physiology of prolonged submaximal exercise and the causes of fatigue, there is a lack

Overload, Performance Incompetence, and Regeneration in Sport, edited by Lehmann *et al.*
Kluwer Academic / Plenum Publishers, New York, 1999.

of understanding of the effects of the marathon or ultra-marathon race on post-race muscle function, damage and regeneration, both in the relative short term (weeks) and in the long term (years). The remainder of this paper will discuss what is known about short and long term recovery after marathon and ultra-marathon running and attempt to identify important research questions relating to these issues.

2. ACUTE EFFECTS OF AN ULTRA-MARATHON

2.1. Depletion of Muscle Glycogen Concentrations

Sherman et al. (1983) showed that immediately after a marathon, muscle glycogen stores were depleted to 40% of pre-race levels in both type I and II fibers (38). Five days later the muscle glycogen levels were still below the pre-race values although glycogen synthase activity was normal. Kirwan et al. (1992) showed that 48 hours after performing exercise which caused muscle damage, subjects had ongoing insulin resistance, with a 37% decrease in insulin-mediated whole body glucose disposal (20). The low muscle glycogen concentrations in the recovery period were attributed to either decreased uptake of glucose through the disrupted sarcolemma in the damaged cells or an increased insulin resistance. An alternative explanation is that there is competition for blood glucose between the inflammatory cells within the muscle fibers and the glycogen depleted damaged fibers (8). The time taken to normalise muscle glycogen concentrations after a marathon is not well known, but is expected to be 10 days or longer (34).

2.2. Skeletal Muscle Damage

Several studies show that muscle damage occurs after marathon and ultra-marathon races. Matin et al., (1983) injected technetium 99m pyrophosphate, which is taken up by damaged cells, into the blood of 11 ultramarathoners after 80 and 160 km races and found that large quantities of this radiolabelled substance appeared in the painful muscles of the subjects' lower limbs (25). Subjects who complained of the most pain after the races had the highest concentrations of technetium 99m pyrophosphate in their painful muscles. Using different methodology, Hidika et al. (1983) showed that severe muscle damage with signs of fiber necrosis and inflammation occurred in muscle biopsies performed on runners after a marathon (18). In a similar study up to 25% of the muscle fibers of runners after a marathon race showed areas of myofibrillar loss (46). Intra- and extracellular edema with endothelial injury, myofibrillar lysis, dilation and disruption of the t-tubule system, and focal mitochondrial degeneration without inflammatory infiltrate was also present in the muscle samples of these runners. Tissues other than the locomotor muscles also show signs of overuse stress after an ultra-marathon race. Matin et al., (1983) showed that in addition to the muscles of the lower limb, the bones of the lower limbs also showed signs of overuse stress (25). A discussion of these side effects are beyond the scope of the paper.

2.3. Cytoskeleton

Structural proteins in skeletal muscle (for example desmin, titin, nebulin) comprise the cytoskeleton and maintain the structural integrity of the myofibrillar lattice (35). Although most research on muscle damage has focused on changes in the contractile proteins actin and myosin, other studies have also implicated structural proteins in the process of muscle damage. A study described longitudinal desmin extensions in muscle samples collected from subjects who had delayed onset muscle soreness (DOMS), suggesting that there was some disruption to the structural integrity of the muscle fibers (15). More recently, this group showed in a rabbit model that a significant amount of desmin was lost in 2.5% of the muscle recruited during 5 minutes of eccentric contractions (24). Disruption of the desmin network in skeletal muscle affects the function of the muscle because these proteins form intermediate filaments which link adjacent Z discs and maintain structural cross-sectional integrity of the muscle fibers (27). Titin is a large structural protein with a chain length of approximately 27000 amino acids and a MW of about 3 million (2). Titin is designed to act as a spring in the muscle by connecting the thick filaments and the Z-disks, preventing the thick filaments from moving from the centre of the sarcomere. Titin's unique structure allows it to accommodate physiological stretch by first straightening without unfolding, and then unfolding a portion of the molecule called the PEVK domain (13). The unfolding of the PEVK portion of the molecule increases the capacity of the muscle to stretch further. The length of the PEVK sequence varies depending on the type of muscle and determines the stiffness of the muscle tissue. For example, the more elastic fast twitch fibers have a higher titin:actin ratio than the less elastic slow twitch fibers (1). Although speculative, muscle damage may cause a reduction of the elastic potential of the muscle by damaging the titin molecule. This hypothesis needs to be studied further. The significance of how the structural proteins, which comprise the cytoskeleton, influence muscle function has been underestimated (47). A potentially greater role of the cytoskeleton on muscle function is supported by the work of Roberts et al., (1997) who showed that the locomotor muscles during horizontal running functioned to hold the "springs" (tendons) rigid so they can store energy with each step (37). Perhaps this ability to store energy with each step may have a greater role in influencing running performance than was previously thought. Similarly, perhaps the extent of damage and regeneration of these structural proteins after a long distance running event may influence the time taken to recover completely.

2.4. Decrement in Performance: Acute and during Recovery

Marathon runners are advised to refrain from racing for 3–6 months between marathons for optimal performance (33). These recommendations are based on anecdotal observations, as there are few scientific studies providing the optimal time for the recovery of muscle function after a marathon. Sherman et al., (1983) showed that leg extensor strength measured isokinetically decreased immediately after a marathon and was not fully recovered after 7 days (38). Chambers et al., (1998) showed that vertical jump height, a measure of leg extensor muscle power, was significantly decreased immediately after a 90 km race, and remained significantly lower than pre-race values for 18 days (6). However, the lack of specificity of the assessments of muscle function in both these studies limits their application to muscle recovery after running a marathon. Based on the findings of Warhol et al., (1985) which showed signs of regeneration in muscle 12 weeks after a marathon, it can be implied that normal muscle function was not fully restored at this stage (46). Changes in neuromuscular function have similarly been reported after a marathon race (30, 31). These data showed that modifications of the neural activation of muscle may occur to compensate for the exercise-induced contractile fatigue which occurs towards the end of a marathon (30, 31). The modification of the neural activation results in an increase in the duration of both the braking and push-off phases in the running stride. There is also an increased EMG activity in all the locomotor muscles, especially during the push-off phase of the running stride (21). These neuromuscular changes may remain for several weeks after a marathon race and account for the decrement in running performance in the recovery phase after a marathon. Skeletal muscles, other than those specifically recuited during running, are also affected after an ultra-endurance race. For example, the endurance capacity of the inspiratory muscles decreased by 27% three days after an 87 km race (19). It is quite likely that this change in the endurance capacity of the inspiratory muscles would negatively effect running performance. At present, however, there are no studies which have systematically examined the time course of changes in running performance immediately after a race until the runner has fully recovered and is able to train and race optimally.

3. CUMULATIVE EFFECTS OF TRAINING AND RACING

3.1. Skeletal Muscle Regeneration

Injured muscle has the capacity to repair and regenerate. Anecdotal observations show that functional skeletal muscle regeneration occurs after an ultra-endurance race, as runners with severe muscular pain for several days after a race make a full recovery and are able to race again after an adequate recovery period (33). These observations are supported by studies which show a progressive repair of the mitochondrial and myofibrillar damage 3 to 4 weeks after a marathon race. After 8 to 12 weeks there are still signs of muscle regeneration including central nuclei and increased satellite cells in the biopsy samples (46). The process of muscle regeneration is initiated by the acute phase response to muscle injury which occurs immediately after the race (44). Mononucleated cells, which usually reside in muscle cells and are quiescent are activated by the muscle injury and subsequently provide a chemotactic signal to circulating inflammatory cells. Neutrophils then invade the injured site and promote inflammation by releasing cytokines that attract and activate additional inflammatory cells (45). Satellite cells located beneath the basal lamina undergo an initial activation reaction which results in the enlargement of the nucleus and an increase in DNA synthesis (4). Neutrophils may also release oxygen-derived free radicals which may further damage cell membranes. There is an increase in circulating macrophages which invade the damaged tissue and remove the tissue debris by phagocytosis (4). After the removal of the damaged muscles fibers, the regeneration of new muscle fibers begins within the remaining intact basal lamina. Studies have shown that in the complete absence of the basal lamina, no muscle regeneration can occur. The population of myoblastic cells established beneath the old basal lamina then develop as in normal myogenesis, resulting in mature fibers with peripheral nuclei (4). Newly regenerated fibers are thinner than normal (5), and are characterised by central nuclei (4). If the skeletal muscle regenerative process is incomplete, as occurs with non-innervation of the new muscle fibers, the regenerating fibers will atrophy. This may explain the loss in muscle mass which occurs with ageing after years of repetitive muscle regeneration (14).

3.2. Morphological Changes with Training

Evidence discussed earlier shows that skeletal muscle is damaged after marathons and ultra-marathons, and that muscle regeneration occurs in the recovery period. However, this leads to many questions, for which at present there are no clear answers. For example, does muscle have a finite ability to adapt and regenerate? Does repetitive muscle damage and regeneration lead to any associated muscle pathology? How does the repetitive muscle damage influence the natural ageing process? Perhaps the years of training are as stressful as the racing event itself. The anecdotal evidence from top class marathon runners suggests that they have about a 10 year period during which they can expect to perform well in their age-group (33). Thereafter, their decline in marathon running performance may be expected to occur at a faster rate than is expected for their age. This anecdotal observation supports the hypothesis that there is cumulative fatigue after several years of training and racing marathons which results in the skeletal muscle "ageing" at a faster rate that is expected. Scientific proof to support this anecdotal observation is emerging. Kuipers et al., (1989) studied runners over a 7 month period while they trained for a marathon (22). They found a gradual increase in degenerative changes in both type I and type II fibers in the subjects' vastus lateralis muscles over this period. They suggested that these pathological changes were minor and were related to the total distance covered in training rather the intensity of training. However, abnormal mitochondria and signs of muscle fiber regeneration and inflammation were found in the "resting" muscle biopsies of experienced marathon runners (17, 18, 46). Sjöström et al. (1988) studied national class runners and found that the overall morphological picture of the marathon runners varied between subjects (41). Only one of the five runners in the study had normal muscle structure. The abnormalities in the muscle biopsies collected before the race included a poor organisation of muscle fascicles, abundant connective tissue and the majority of fibers showed one or more central nuclei. Other abnormalities included flat angular fibers and signs of fiber type grouping. Both these signs are pathognomonic of denervation atrophy (12, 42) and suggest incomplete regeneration. Sjöström et al., (1988) suggested that the type II fibers are more vulnerable to damage as the two runners in their study with the lowest numbers of type II fibers had the least muscle pathology (41). The suggestion of increased vulnerability of type II fibers is derived from another report in which a well trained 46 year old runner ran 3529 km in 7 weeks (40). The relative amount of type I fibers was higher after the 7 weeks, suggesting either that the number of type II fibers had been reduced after the long run, or the number of type I fibers had increased. This may be a simple training effect. However, the biopsy samples in this study were also characterised by fibers of varying sizes and an increase in central nuclei in approximately 30% of the fibers. But there are examples in the literature of experienced runners with no apparent abnormalities for their age. For example, Maud et al., (1981) studied a 70 year old, previously world class long distance runner, who had been training for 52 years (26). He was still training, had no orthopaedic abnormalities and apart from thickened heel pads was declared healthy. Unfortunately no data are available on his muscle morphology. In addition a recent study of the top 10 finishers in each group of a 56 km race showed that some of the runners who were the top performers in the 60 year old category had been running for up to 30 years, in contrast to other runners in this group who had recently started running (23). This raises interesting questions. Do the runners who had been running for 3 decades represent runners who are biomechanically perfect, or do these runners have muscle and connective tissue which is resistant to the long term changes which have been described above? These are questions which need to be answered because they may influence the recommendations for the training of marathon runners in the future.

3.3. Skeletal Muscle Pathology Associated with High Volume Training

St Clair Gibson et al. (1998) reported on a 28 year old runner who experienced a rather sudden decline in running performance and an inability to tolerate high training loads (43). The subject began running at the age of 12 years and by the age of 19 years was the national junior champion with a best 10 km time of 28:35 (min:s). The subject was accustomed to high volume, high intensity training (about 6000 km per year). A vastus lateralis muscle biopsy revealed a type I fiber predominance with no signs of inflammation, necrosis or excessive regeneration. The mitochondria, however were grossly abnormal, varied in size and contained a dense matrix with an increased number of coarse and broad cristae. The abnormal mitochondria were observed in large subsarcolemmal aggregates (Figure 1). A muscle biopsy, taken from the triceps muscle, was completely normal (Figure 2). The authors concluded that the mitochondrial myopathy in the vastus lateralis either; (i) existed previously but was undiagnosed, (ii) was acquired as a result of unknown agents, or (iii) was acquired after prolonged training. The fact that the myopathy was located to the lower limbs suggests that this was not a classical mitochondrial myopathy and was caused by either (ii) or (iii).

Another example of this mitochondrial pathology associated with excessive exercise was that of Greg Le Mond, who won the Tour de France three times, and the world championship twice (11). The "red herring", however, in his case was that he was involved in a shooting accident and had lead pellets imbedded in his body which may have caused the myopathy. Derman et al., (1997) have described a clinical condition, the fatigued athlete myopathic syndrome (FAMS) (10). The symptoms of this syndrome varied but the common features of athletes with this condition were; (i) they had a history of high volume training for many years, (ii) they presented with chronic fatigue and decreased exercise performances and a clinical picture which was dominated by skeletal muscle symptoms including excessive DOMS, stiffness, tenderness and skeletal muscle cramps, and (iii)

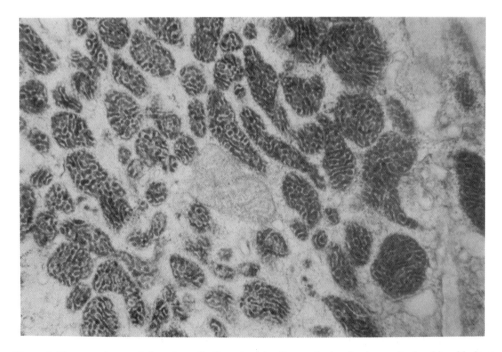

Figure 1. Electron micrograph from the patient's vastus lateralis muscle showing large mitochondria with dense matrices and coarse, abnormal cristae. A normal mitochondria is visible in the center of the figure (original magnification X35 040) (reference 43).

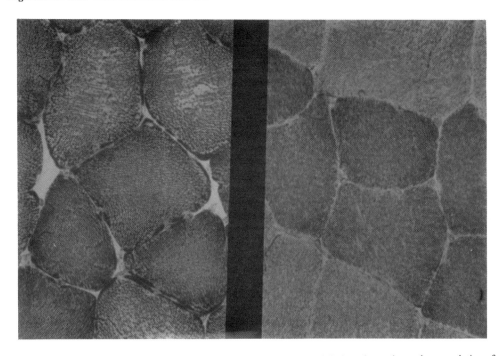

Figure 2. NADH stain of patient's vastus lateralis muscle (left) showing lobular subsarcolemmal accumulation of the oxidative enzyme (original magnification, X400). NADH stain of patient's triceps muscle (right) showing a fairly even distribution of the oxidative enzyme throughout the fibers; type I fibers are more intensely stained than type II fibers (original magnification X400) (reference 43).

they had often consulted many clinicians unsuccessfully. Some of the pathology described in the patients in this study included irregular muscle fiber size and distribution, and an alteration in mitochondria structure with lipid inclusions. Interestingly, 6 of the 8 subjects studied had evidence of previous Ebstein Barr virus infections.

4. SUMMARY AND CONCLUSIONS

Much research has addressed the causes of fatigue arising during ultra-endurance marathon running. It is known that this fatigue has several origins, the most likely being muscle glycogen depletion, hypoglycaemia, brain neurotransmitter changes, neuromuscular changes and CNS effects associated with overheating. However, less is known about the acute effects an ultra-endurance marathon has on muscle and running performance, how muscle regenerates after severe damage, and whether repetitive skeletal muscle damage induced by long term training and racing induces any chronic pathological and morphological changes. Studies performed in the 1980's showed that runners had signs of pathology in their muscles after a race. This muscle damage was characterised by myofibrillar loss, intra- and extracellular edema and disruption of the t-tubule system. Evidence of progressive repair of the mitochondrial and myofibrillar damage were visible 3 to 4 weeks after the marathon with signs of regeneration in the muscle still evident after 12 weeks. These scientific data tend to support the anecdotal observations that marathon runners need about 3 or more months to recovery from a marathon race. Recent data show that muscle and connective tissue units function as springs which absorb the forces on impact with each stride. However, the elastic properties of muscle have not been well studied. Many structural proteins in muscle have been identified but their function and contribution to the cytoskeleton is not well understood. However, as more information becomes available about the cytoskeleton in skeletal muscle, it is evident that the structural proteins, desmin and titin in particular, have an important role in maintaining the myofibrillar lattice, and in contributing to the elastic properties in muscle. The maintenance of structural integrity is important for optimal muscle function, and is disrupted when muscle is damaged. Loss of desmin from skeletal muscle is the first measurable marker of muscle damage, suggesting that it may be the "weak link" in muscle structure. Titin is a unique protein molecule which has special properties of elasticity and acts as an elastic connector between the myosin filaments and the Z-disks and is assumed to be functionally important. How the properties of titin are affected when muscle is damaged and how it regenerates after injury is not well understood. This needs to receive attention in the future for a better understanding of the loss of forces accompanying muscle damage. Studies show that training for marathons may be as physically stressful as the race itself. The literature has many examples of experienced runners with skeletal muscle pathology, even in the "rested" state, suggesting that there is some cumulative effect of years of training. Whether these changes in the skeletal muscle are merely a form of chronic adaptation or whether they have pathophysiological manifestions remains to be determined. However, the anecdotal observations are that runners have a period of time during which they can adapt to training and perform well, followed by a difficulty in sustaining high training loads which coincides with a decline in running performance. Often this decline in performance occurs at a rate which exceeds that rate which can be accounted for by the normal ageing process. This observation supports the fact that the changes in the skeletal muscle are pathological. This raises several questions. Why are some runners more vulnerable to the development of muscle pathology than others? Is the vulnerability to muscle pathology related to poor biomechanics which results in the athletes having to sustain higher impact stresses when they train? Do their muscles lack the ability to fully regenerate after an exposure to damaging exercise? These questions need to be answered before the short and long term effects of marathon running are fully understood. This information will determine the training advice given to marathon runners of the future.

REFERENCES

1. Askter HA, Granzier HlM and Focant B (1989) Differences in the I band structure, sarcomere extensibility, and electrophoresis of titin between two muscle types of the perch (Percafluviatilis l). J Ultrastruct Mol Struct Res 102: 109–121.
2. Baringa M (1995) Titanic protein gives muscles structure and bounce. Science 270: 236.
3. Bosch AN, Dennis SC and Noakes TD (1993) Influence of carbohydrate loading on fuel substrate turnover and oxidation during prolonged exercise. J Appl Physiol 74: 415–423.
4. Carlson BM (1995) Factors influencing the repair and adaptation of muscles in aged individuals: satellite cells and innervation. J Gerontol A Biol Sci Med 50: 96–100.
5. Carlson BM and Faulkner JA (1983) The regeneration of skeletal muscle fibers following injury: a review. Med Sci Sports Exerc 15: 187–198.
6. Chambers C, Noakes TD, Lambert EV and Lambert MI (1998) Time course of recovery of vertical jump height and heart rate vs running speed after a 90 km foot race J Sports Sci 16:645–651.
7. Costill DL (1979) A scientific approach to distance running. Track and field News, 1979 Los Altos, California, USA.
8. Costill DL, Pascoe DD, Fink WJ, Robergs RA, Barr SI and Pearson D (1990) Impaired muscle glycogen resynthesis after eccentric exercise. J Appl Physiol 69: 46–50.

9. Davies CTM and Thompson MW (1986) Physiological responses to prolonged exercise in ultramarathon athletes. J Appl Physiol 61: 611–617.

10. Derman W, Schwellnus MP, Lambert MI, Emms M, Sinclair-Smith C, Kirby P, and Noakes TD (1997) The "worn-out athlete": A clinical approach to chronic fatigue in athletes. J Sports Sci 15: 341–351.

11. Derr M (1995) The end of the road. Is a new malady afflicting elite athletes? Scientifica America April, 10 - 1.

12. Dubowitz V (1985) Muscle biopsy: a practical approach. Bailliere Tindall Eastbourne, England, 2nd ed.

13. Erickson HP (1997) Stretching single protein molecules: titin is a weird spring. Science, 276: 1090–1092.

14. Faulkner JA and Brooks SV (1995) Muscle fatigue in old animals. Unique aspects of fatigue in elderly humans. Adv Exp Med Biol 384: 471–480.

15. Fridén J, Kjörell U and Thornell L-E (1984) Delayed muscle soreness and cytoskeletal alterations: an immunocytological study in man. Int J Sports Med 5: 15–18.

16. Gollnick PD, Korge P, Karpakka J and Saltin B (1991) Elongation of skeletal muscle relaxation during exercise is linked to reduced calcium uptake by the sarcoplasmic reticulum in man. Acta Physiol Scand 142: 135–136.

17. Goodman C, Henry G, Dawson B, Gillam I, Beilby J, Ching S, Fabian V, Dasig D, Kakulas B and Morling P (1997) Biochemical and ultrastructural indices of muscle damage after a twenty-one kilometre run. Aus J Sci Med Sport 29: 95–98.

18. Hikida RS, Staron RS, Hagerman FC, Sherman WM and Costill DL (1983) Muscle fiber necrosis associated with human marathon runners. J Neuro Sci 59: 185–203.

19. Ker JA and Schultz CM (1996) Respiratory muscle fatigue after an ultra-marathon measured as inspiratory task failure. Int J Sports Med 17:493–496.

20. Kirwan JP, Hickner RC, Yaresheski KE, Kohrt WM, Wiethip BV and Holloszy JO (1992) Eccentric exercise induces transient insulin resistance in healthy individuals. J Appl Physiol 72: 2197–2202.

21. Komi PV, Hyvärinen T, Gollhofer A and Mero A (1986) Man-shoe-surface interaction. Special problems during marathon running. Acta Univ Oulu, 179: 69–72.

22. Kuipers H, Janssen GME, Bosman F, Frederik PM and Geurten P (1989) Structural and ultrastructural changes in skeletal muscle associated with long-distance training and running. Int J Sports Med 10 (Suppl 3): S156-S159.

23. Lambert MI and Keytel L Training habits of top male and female Two Oceans runners. Two Oceans Marathon Programme, 1998.

24. Lieber RL, Thornell L-E and Fridén J (1996) Muscle cytoskeletal disruption occurs within the first 15 min of cyclic eccentric contraction. J Appl Physiol 80: 278–284.

25. Matin P, Lang G, Carretta R, Simon G (1983) Scintigraphic evaluation of muscle damage following extreme exercise: concise communication. J Nucl Med 24: 308–311.

26. Maud PJ, Pollock ML, Foster C, Anholm JD, Guten G, Al-Nouri M, Hellman C and Schmidt DH (1981) Fifty years of training and competition in the marathon: Wally Hayward, age 70 - a physiological profile SAMJ 59: 153–157.

27. McComas J (1996) Skeletal muscle - form and function Human Kinetics Champaign IL USA.

28. Newsholme E A, Blomstrand E and Ekblom B (1992) Physical and mental fatigue: metabolic mechanisms and importance of plasma amino acids. Br Med Bull 48: 477–495.

29. Newsholme EA, Leech T and Duester G (1994) Keep on running—the science of training and performance John Wiley and Sons, New York.

30. Nicol C, Komi PV and Marconnet P (1991a) Fatigue effects of marathon running on neuromuscular performance I Changes in muscle force and stiffness characteristics. Scand J Med Sci Sports 1: 10–17.

31. Nicol C, Komi PV and Marconnet P (1991b) Fatigue effects of marathon running on neuromuscular performance II Changes in force, integrated electromyographic activity and endurance capacity. Scand J Med Sci Sports 1: 18–24.

32. Nielsen B, Strange S, Christensen NJ, Warberg J and Saltin B (1997) Acute and adaptive responses in humans to exercise in a warm, humid environment. Pflügers Archives 434: 49–56.

33. Noakes TD (1992) Lore of Running. Oxford University Press, Cape Town.

34. O'Rielly KP, Warhol MJ, Fielding RA, Frontera WR, Meredith CN and Evans WJ (1987) Eccentric-induced muscle damage impairs muscle glycogen repletion. J Appl Physiol 63: 252–256.

35. Patel TJ, Lieber RL (1997) Force transmission in skeletal muscle: from actomyosin to external tendons. Ex Sports Sci Rev 25: 322–363.

36. Rauch HG, Hawley JA, Noakes TD and Dennic SC (1998) Fuel metabolism during ultra-endurance exercise Pflügers Archives 436:211–219.

37. Roberts TJ, Marsh RL, Weyand PG, Taylor, CR (1997) Muscular force in running turkeys: the economy of minimising work. Science, 275: 1113–1115.

38. Sherman WM, Costill DL, Fink WJ, Hagerman FC, Armstrong LE and Murray TF (1983) Effect of a 42.2 km footrace and subsequent rest or exercise on muscle glycogen and enzymes. J Appl Physiol 55: 1219–1224.
39. Sjödin B and Svedenhag J (1985) Applied physiology of marathon running. Sports Med 2: 83–99.
40. Sjöström M, Fridén J and Ekblom B (1987) Endurance, what is it? Muscle morphology after an extremely long distance run. Acta Physiol Scand 130: 513–520.
41. Sjöström M, Johansson C and Lorentzon R (1988) Muscle pathomorphology in m. quadriceps of marathon runners. Early signs of strain disease or functional adaptation? Acta Physiol Scand 132: 537–542.
42. St Clair Gibson A (1997) Neural and humoral control of muscle atrophy. PhD Thesis, University of Cape Town.
43. St Clair Gibson A, Lambert MI, Weston AR, Myburgh KH, Emms M, Kirby P, Marinaki AM, Owen EP, Derman W and Noakes TD (1998) Exercise-induced mitochondrial dysfunction in an elite athlete. Clin J Sports Med 8: 52–55.
44. Strachan AF, Noakes TD, Kotzenberg G, Nel AE and De Beer FC (1984) C-reactive protein levels during long-distance running. Br Med J 289: 1249–1251.
45. Tidball JG (1995) Inflammatory cell response to acute muscle injury. Med Sci Sports Exerc 27: 1022–1032.
46. Warhol MJ, Siegel AJ, Evans WJ and Silverman LM (1985) Skeletal muscle injury and repair in marathon runners after competition. Am J Pathol 118: 331–339.
47. Waterman-Storer CM (1991) The cytoskeleton of skeletal muscle: is it affected by exercise? A brief review. Med Sci Sports Exerc 23, 1240–1249.

14

NEUROENDOCRINE SYSTEM

Exercise Overload and Regeneration

A. C. Hackney

Endocrine Section – Applied Physiology Laboratory
University of North Carolina
Chapel Hill, North Carolina

1. INTRODUCTION

The physiological systems of the human body are placed under a great deal of stress during the process of exercise training. Within elite athletes this stress can be of enormous levels because of the volume and intensity of work they perform in their training regimes (4, 13, 29, 104). When training stress is of an appropriate level then there is a positive adaptation in the human organism and typically physical performance improves. However, the converse can also be true—inappropriate levels of stress can result in negative, maladaptations and declines in physical performance (4, 104). In the adaptive process to exercise training, it is well known that there is a need for rest and recovery from the stress of training. If this is not allowed, repetitive exposure to what would normally be an appropriate level of stress can become inappropriate in nature (4, 75, 104). When an inappropriate level of training stress is imposed upon the athlete there is a great likelihood of the athlete developing the "overtraining syndrome". Once this occurs, physical performance can decline so severely that the athlete's competitive season may be over (41).

How much rest and recovery is necessary in a training program is a question of extensive debate among coaches, athletes, and sports medicine scientists. This is a complicated issue for several reasons. First, athletes are very individualistic in the physiologic adaptation process because of genetic factors, age, gender, and prior training history (4, 13, 78, 84, 92, 104). Secondly, research evidence suggests that the amount of time necessary for the recovery from the stress of training may vary within the different organ systems of the human body (e.g., cardiovascular vs. musculoskeletal) (4, 103, 104). Thirdly, the training cycles of athletes are variable and complex, involving periods of "reduced-load", "overload" and "overreaching" type training (see references 4, 13, and 64 for discussion). This may result in the need for varying amounts of rest and recovery depending upon the training cycle the athlete is currently performing in their program. Finally, psychologically, athletes have wide variances in the extent to which they can tolerate the

Overload, Performance Incompetence, and Regeneration in Sport, edited by Lehmann *et al.*
Kluwer Academic / Plenum Publishers, New York, 1999.

physiological and emotional stress of training (75). This variability in their psychological demeanor can impact upon the amount of rest and recovery they need with training.

The intent of this paper is to review the research that has dealt with one physiological system in particular and to focus on the issue of rest and recovery within that system. The system to be addressed is the neuro-endocrine system. The specific aim of the discussion on this topic is to examine how much rest and recovery time may be necessary for the restoration of the neuro-endocrine system from the stress of exercise. It is hoped that this paper can serve as a preliminary step in the development of guidelines as to the amount of rest and recovery necessary to prevent a negative, mal-adaptation in the neuro-endocrine system in response to the stress of exercise training.

The organizational presentation of investigative findings in this paper is structured into research studies examining responses for: *a*) acute-single bouts of exercise, *b*) chronic exercise training exposure, and *c*) mechanisms of potential mal-adaptation.

2. DEFINITION OF TERMS AND DELIMITATIONS

In order to make the discussion within this paper as lucid as possible, several working definitions are necessary.

Neuro-endocrine system is defined as the autonomic nervous system and the traditional, classic endocrine glands (e.g., pituitary, thyroid, adrenal) and does not encompass the hormonal-like actions of tissues that are not anatomical defined as endocrine glands (e.g., liver, atrium). The responses of the neuro-endocrine system are quantified based upon the hormonal changes observed in the blood.

Exercise is used to refer to an "acute bout of physical activity" and the phrase *exercise training* deals with exposure to "chronic, regimented programs of physical activity".

Rest-recovery is defined on several levels; *a*) in reference to the amount of time between exercise sessions (i.e., workouts) on a daily or weekly basis, and *b*) in reference to the amount of time between longer cycles or periods of exercise training. Relative to this last definition, the term *regeneration* is sometimes used in European literature to refer to cycles or periods of extended rest-recovery within a training regime. In this paper, the term regeneration will be used in this context.

It is also necessary to set certain delimitations to the scope of this review to allow the paper to remain focused and concise. First, only neuro-endocrine research dealing with humans is addressed. Secondly, factors such as aging and gender differences (men versus women as well as menstrual cycle influences in women) are not discussed. Finally, the discussion is centered on the hormonal responses from only select endocrine glands considered essential to the physiological functioning during exercise. The reader is directed to several excellent review articles for more in-depth discussion of the complete endocrinological responses to exercise and other stresses (89, 92, 99, 100).

3. ACUTE-SINGLE BOUTS OF EXERCISE

3.1. Daytime vs. Nighttime Responses

The hormonal responses to exercise are affected by the intensity and duration of an exercise bout. Research indicates that most hormonal changes (i.e., increases or decreases) are approximately proportional to the intensity and, or duration of an exercise bout (14, 21, 29,

34, 57, 58, 62, 64, 73, 93, 103). The absolute and relative magnitude of change in individual hormones in response to exercise can be rather dramatic (103). But, most evidence would suggest that these changes are very transient in nature (29, 31, 103). For example, Hackney and associates studied several hormonal indices (cortisol, prolactin, growth hormone, testosterone, luteinizing hormone, follicle-stimulating hormone, and sex-hormone binding globulin [SHBG]) during the recovery from different types of exercise. In this study subject hormonal profiles were examined on three separate days. One was a control day with no exercise, another was a day with 60 minutes of aerobic running at 70% maximal oxygen uptake (VO_{2max}), and the third was a day with anaerobic intervals (2 minutes exercise at ~110% VO_{2max}, 2 minutes rest) for ~60 minutes (42). In this study the exercise took place during the first hour of blood sampling (starting at 0700 h) and the total volume of work performed during the two exercise bouts was of an equal amount. Hormones were measured hourly for 9 consecutive hours on each of the days. The results demonstrated that following an acute bout of either aerobic or anaerobic exercise (i.e., representative of typical training sessions) that most disturbances in the neuro-endocrine system are during the early recovery phase (≤ 1 to 4 hours postexercise). Similar findings have been reported by many investigators for a variety of hormones, although most studies have sampled hormonal responses for shorter periods of time into recovery (14, 34, 29, 74, 77, 80, 81, 83, 90, 91, 106).

Notwithstanding, if hormonal observations are extended into the nighttime (i.e., nocturnal) then different responses seem to exist. That is, the disturbances of hormonal concentrations at night following daytime exercise appear to be of greater magnitude and duration than those just reported above. Hackney et al. examined hourly hormonal responses over a 12-hour nighttime period (2000 h to 0800 h) on two separate days (43). One day was a control day in which no exercise was performed, and the other day was one in which 90 minutes of aerobic exercise at 70% VO_{2max} was performed (running from 1630 h to 1800 h). Summaries of the results from this study are shown in Table 1. The results indicated that growth hormone and cortisol are significantly suppressed at night after daytime exercise while prolactin and thyroxine display the opposite responses. The German researchers, Kern et al. (54) have published very similar findings concerning nighttime hormonal responses. This latter work also suggests that the intensity of daytime exercise plays a role in the degree to which nighttime disturbances develops in hormonal concentrations. That is, the disturbance effect seems somewhat proportional to the magnitude of the daytime exercise. Another study that examined hormonal concentrations at night was conducted by Wittert and associates (110). These investigators studied ultra-marathon runners during their recovery (3–5 days afterward) from a 15 hour competition. Blood adrenocorticotropic hormone (ACTH) and cortisol concentrations over a 16 hours period

Table 1. The integrated nocturnal hormonal response (mean±SE) of men over a 12 hour period from 20:00 h (8:00 PM) until 08:00 h (8:00 AM) the next day on a control day (involving no daytime exercise) and on an exercise day (involving 90 minutes continuous activity at 70% maximal oxygen uptake)[a]

Hormone	Measurement units	Control condition	Exercise condition	Probability level
Testosterone	$ng \cdot ml^{-1} \cdot h^{-1}$	106.4 ±15.8	115.7 ±14.8	0.95
Luteinizing hormone	$mIU \cdot ml^{-1} \cdot h^{-1}$	58.5 ±4.4	68.7 ±7.8	0.54
Prolactin	$ng \cdot ml^{-1} \cdot h^{-1}$	171.1 ±12.2	338.5 ±15.8	<0.01
Growth hormone	$ng \cdot ml^{-1} \cdot h^{-1}$	37.1 ±15.9	10.5 ±1.9	0.05
Thyroxine	$\mu g \cdot dl^{-1} \cdot h^{-1}$	99.5 ±2.2	108.1 ±2.7	<0.01
Cortisol	$\mu g \cdot dl^{-1} \cdot h^{-1}$	164.7 ±18.7	76.2 ±12.8	<0.01

[a]See Hackney et al. reference (43).

that included nighttime sampling, as well as a 24-h urinary cortisol measure was examined in these runners. The responses of the ultra-marathon runners were compared to control subjects who had not exercised. Results showed there were no main differences for cortisol in the blood or urine between the groups; but blood ACTH levels were elevated in the ultra-marathon runners. The authors interpret their findings as indicative of the development of an adrenal cortex resistance to central signs (i.e., ACTH) in the ultra-marathon runners (77, 105). This study is however different than the others just mentioned above because it involved exercise that was a competition and not a training session. Many studies have been conducted on exercise that is of a "competition-nature" as opposed to exercise of a "training-nature". Such studies have shown that it is not uncommon for hormonal disturbances to exist for 24, 48, and 72 h following competitive events (e.g., a 42 kilometer running marathon, triathlon, or similar event) (2, 22, 63, 76, 82). Such findings have been interpreted as indicating that competitions are much more stressful than exercise training on the neuro-endocrine system (30, 31, 38, 39). However, it should be noted that in many of these "competition" studies, blood-sampling frequency has been extremely limited (e.g., every 12 to 24 hour) which makes the data somewhat suspect and difficult to interpret (11, 28). More research is necessary to study this issue carefully and determine the degree of recovery necessary following a competitive event.

It also appears that the mode of activity has an effect on the nocturnal neuro-endocrine responses to an acute-single bout of exercise. McMurray et al. (72) has examined nighttime hormonal responses following a daytime resistance exercise bout. In this work growth hormone, cortisol and thyroxine were unaffected by the daytime resistance exercise. However, testosterone and triiodothyronine were significantly elevated during the night following the daytime exercise. These hormonal results are somewhat contradictory to those just presented above on running exercise sessions (see reference 42). Perhaps these differences are due to a need for varying lengths of recovery time between exercise sessions involving different activity modes (77, 81, 99, 101). The need for various lengths of recovery may be because the time course of peripheral tissue-specific adaptations may differ. This concept needs to be addressed specifically in future research studies, but is supported to some degree by other investigators (2, 25, 29, 32, 39, 59, 60, 70, 103, 104).

An additional point to consider is multiple daily training sessions. Our research group has looked at hormonal concentrations over a 24-h period on a control day with no exercise, on a day with aerobic exercise (running) in the morning and afternoon, and on a day with anaerobic intervals in the morning and afternoon (Hackney unpublished findings). Preliminary findings from this work indicate that; *a*) immediately following an exercise bout (in the morning or afternoon) there are hormonal disturbances that are relatively transient in nature, and *b*) hormonal concentrations at night after daytime are profoundly different than on a night with no daytime exercise. As of yet, it can not be determined whether multiple daily training sessions, or anaerobic versus aerobic exercise place a greater stress upon the neuro-endocrine system and thus may require more time for rest-recovery (23). These latter issues require further experimental study, which we are currently undertaking. Notwithstanding, related research by Costill and associates (16) indicates that twice-daily training sessions may not allow adequate recovery time for athletes and therefore may not result in improved physical performance.

3.2. Summary

Collectively, the findings of the above studies could be interpreted as suggesting that the recovery of the neuro-endocrine system may not be complete within the 24 hours following an exercise bout. However, what is unclear is whether the changes observed might be responses that are allowing the organism to adapt in a positive fashion to the stimulus of exercise, or if they represent a deviant response reflective of excessive stress upon the system. It is critical in the immediate future that researchers investigate this last point.

4. CHRONIC EXERCISE TRAINING EXPOSURE

4.1. Experimental Studies—Overload Periods

To some degree, all athletic training programs incorporate the basic fundamental training principle of "progressive resistance exercise" or "overload" (13). This approach can typically involve several "back-to-back" days of strenuous exercise; or, periods of weeks involving an increased volume – intensity of training (i.e., an "overload" training cycle) (64). This type of hard, consecutive training has the potential to place great strain upon the neuro-endocrine system. Several research groups have tried to experimentally re-produce such training under laboratory settings, and to quantify the degree of hormonal responses. An example of this is the research work of Fry and associates (26). In this study male subjects did intensive-interval training (running) for 10 days followed by 5 days of recovery involving minimal physical activity. The hormone measures were evaluated in the resting state and consisted of testosterone, cortisol, growth hormone, and SHBG. Changes occurred in nearly all these measures during the training period ($\sim \pm 10\%$ to 25% changes in concentration levels from pre-exercise baseline). During the 5 days of recovery, however, more substantial hormonal changes were noted ($\sim \pm 25\%$ to 45% changes in concentration levels from pre-exercise baseline). Kirwan et al. (56) has reported similar findings for swimmers over an 11 day period of intensive, hard training. Unfortunately, these latter investigators did not examine physiological responses during any recovery days following the intensive training. Somewhat in contrast to these findings, Shepley et al. (85) and Houmard et al. (52) have studied athletes over several weeks of heavy training followed by periods of reduced training (i.e., a "taper period", in which the subjects were allowed partial rest and recovery). These investigators indicate that a taper period follow-ing hard training results in improved cardiovascular function (i.e., recovery). However, no substantial hormone changes were apparently noted during the training, and the taper pe-riod had minimal effects on the hormonal levels when compared to the training period. Yet, other investigators have reported that if intensive training is extended for long periods of consecutive days (2–4 weeks) then resting hormone levels can change very dramati-cally with relative changes in concentrations of approximately 50% to 100% having been reported (29, 31, 49, 101, 107, 108). Perhaps these conflicting findings reflect differences in the initial level of training of the subjects, which can impact upon resting hormonal re-sponse (8, 35, 37, 92, 94, 107). Furthermore, in the "taper studies" possibly the taper pe-riod was insufficient in the amount of rest-recovery incorporated to allow a restoration of the neuro-endocrine system.

4.2. Athletic Training Studies

Several research groups have attempted to very closely mimic the actually micro- and macro-training cycles that athletes typical incorporate into their training. Urhausen and associates (93, 94, 95, 96, 97) have conducted several such studies. One of the pri-mary hormonal findings of these studies is that testosterone (total and free) are reduced during periods of heavy training, while cortisol levels have a tendency to be elevated. These changes result in a significant decline in the circulating testosterone:cortisol (T:C) ratio. Conversely, when periods of rest-recovery-regeneration are incorporated into the training cycle the ratio is elevated. Alen et al. (2), Bonifazi et al. (9) and Fellmann et al. (24) report very similar hormonal responses to such training cycles. This series of hormonal changes has been interpreted as an indication that the body is shifting back and forth between a more catabolic (i.e., decrease ratio) and anabolic (i.e., increased ratio) hormonal status (6, 15, 55, 61, 82, 86, 87, 94, 102). Numerous investigations in the literature have used the T:C ratio as a proposed marker of the anabolic versus catabolic status in athletes undergoing training (2, 26, 64, 87, 88, 93). However, other investigators have argued that this may be an overly simplified interpretation of this hormonal measurement (25, 39, 60, 64, 84). Regardless of its precise physiological meaning, this ratio measure does seems to be a rather consistent responder to the stress of exercise training Therefore it may provide some degree of insight as a "stress reactivity" index in athletes.

Lehmann and associates (69) have measured nocturnal urinary catecholamine responses for soccer players during the competitive season when heavy training or competition was occurring and during a winter break in the season when a period of regeneration was allowed (i.e., several days of rest). The catecholamine excretion levels during heavy training and competition are substantially lower than during the regeneration period. This same change was also reported by Lehmann and associates in cyclists preparing for the Olympic Games (66, 68) and by Nassens et al. (see reference 68) in another group of soccer players during their season. Yet, other research indicates that plasma catecholamine levels tend to show the opposite responses of urinary catecholamines. Hooper et al. (51) reported in swimmers that plasma norepinephrine level is higher during intensive training as opposed to periods of reduced training ("tapering") or a recovery period (i.e., regeneration days). Lehmann et al. (68) has also shown elevated plasma norepinephrine and epinephrine at rest as well as in response to submaximal exercise in track athletes. This elevation in the catecholamines occurred in these athletes when they were doing high-volume, intensive training versus periods of lower volume training.

At first these contrasting urinary and plasma catecholamine results just noted above seem paradoxical in nature. However, the following explanations can be offered. First, nocturnal urinary catecholamine levels are thought to be reflective of basal conditions and thus are an indicator of intrinsic sympathetic activity—therefore, chronic training without a period of rest-recovery may be lowering the basal activity level of the sympathetic system. Second, plasma catecholamines are thought to be more reflective of acute, stress-related sympathetic responses—therefore, the plasma levels may be elevated because a reduced sensitivity to catecholamines has developed from chronic exercise training without appropriate rest-recovery or regeneration (33, 64, 65, 68). This last point seems feasible and is supported by the findings of Jost and associates (53). These investigators reported decreased beta-adrenergric receptor density in runners and swimmers during heavy training as compared to periods of reduced training.

4.3. Non-Athletic Exercise Studies

The researchers Aakvaag et al. (1) and Opstad et al. (79) have studied military recruits who have undergone 4 to 5 day field maneuvers that involved high levels of physical activity and emotional stress. Their studies have shown tremendous reductions in many circulating hormone levels (e.g., cortisol, testosterone, prolactin, and thyroxine) with the increased strain of these maneuvers. Furthermore, after completion of the maneuvers the

Table 2. The mean (±SE) resting thyroid hormone levels in men immediately before and after a 14-d mountain climbing expedition (peak altitude ~6000 M)[a]

Hormone	Measurement unit	Before expedition	After expedition	Normal basal
Thyroid- stimulating hormone	uIU· ml^{-1}	1.76 ± 0.12	0.82 ± 0.06*	1.63 ± 0.14
Total thyroxine	nmol· l^{-1}	150.1 ± 6.5	127.5 ± 11.5	138.1 ± 3.9
Free thyroxine	pmol· l^{-1}	15.1 ± 0.4	15.0 ± 0.7	15.2 ± 0.5
Total triiodothyronine	nmol · l^{-1}	3.9 ± 0.2	3.0 ± 0.2*	3.8 ± 0.2
Free triiodothyronine	pmol · l^{-1}	3.8 ± 0.3	2.6 ± 0.1*	4.3 ± 0.2

[a]Values for these men are also reported for a basal period in which the men were completely rested. The * denotes values that are significantly (p<0.01) reduced from the before expedition and basal values. See Hackney et al. (45) reference.

recovery of normal basal levels was extremely slow, taking days or weeks (depending upon the magnitude of the hormonal change during the maneuvers and each select hormone). Hackney et al. (45, 47, 48) have examined military personal during field maneuvers over longer periods of time (one to two weeks) and found similar results. The longest period of time that these investigators studied (14 days) involved a mountain climbing expedition. Hormonal responses were observed in men before an expedition to climb to an altitude of 6000m, and immediately after the 14-day expedition. Table 2 reports select responses from this study. There was a marked suppression of nearly all the thyroid hormones in the subjects. Such hypo-thyroidal states are extremely counter-productive physiologically as they are associated with a compromised mental and muscular function - performance (7, 19, 36, 46, 109). Obviously, this and the other field-based military studies mentioned have some confounding factors that make direct application of the results to athletes somewhat difficult. Nonetheless, these studies provide some degree of insight into the workings of the neuro-endocrine system when it is under stress.

4.4. Summary

The above research findings suggest that when physical workload and emotional stress are excessive without any degree of regeneration, severe hormonal disturbances can occur which could be very counter-productive to the physiological recovery of other tissues and organs in the body. The implications, relative to athletes and coaches, from these studies might be that exercise training should not include periods of to many consecutive, hard training days.

5. MECHANISM INVESTIGATIONS

Many studies have reported with excessive levels of chronic, hard exercise training that abnormal hormonal changes occur that could impact effect certain physiological functions of the body in a negative fashion (18, 27, 39, 41, 64, 65, 66, 68). The physiological mechanisms that might be bringing about these alterations in the neuro-endocrine function are unclear, but a few studies have used research protocols to try to discern the problem.

In one such study, Barron et al. (5) studied ultra-marathon runners who had become overtrained. In this study there was the use of a drug challenge (iv injection) to stimulate the pituitary release of growth hormone, thyroid-stimulating hormone, prolactin, luteinizing hormone, follicle-stimulating hormone, ACTH, and cortisol from the pituitary and ad-

Table 3. The mean (±SE) integrated hormonal responses (90 min) above basal levels after an IV insulin injection in male marathoners displaying overtrained characteristics, after 4-wk rest in the same marathoners, and compared to a control group of non-overtrained marathoners[a]

Hormonal assessment	Measurement Unit	Marathoners (overtrained)	Marathoners (4-wk rested)	Marathoners (controls)
Growth hormone	$(ng \cdot min \cdot ml^{-1}) \times 10^{-3}$	$0.9 \pm 0.4^*$	2.0 ± 0.3	2.1 ± 0.2
ACTH	$(pg \cdot min \cdot ml^{-1}) \times 10^{-3}$	$0.3 \pm 0.2^*$	2.6 ± 1.4	4.3 ± 1.5
Cortisol	$(nmol \cdot min \cdot l^{-1}) \times 10^{-3}$	$2.0 \pm 2.3^*$	15.6 ± 1.8	12.7 ± 3.2

[a]The * denotes that overtrained marathoner responses were significantly ($p<0.05$) less than the other values. There was no significant difference between the responses of the overtrained marathoners after 4-wk rest and the responses in the control group. The values reported are from the Barron et al. (5) reference and have been mathematically adjusted to the nearest one tenth of a value.

renal cortex, respectively. The challenge was administered when the ultra-marathoners were diagnosed as being severely overtrained, and again after a 4-week recovery period (see references 41, 65, 75, and 87 for symptom overview). Additionally, the challenge was performed in a group of runners who were not overtrained (therefore, serving as a control group). Partial results from this study are displayed in Table 3. The growth hormone, ACTH and cortisol release were greatly suppressed in the marathon runners when they were diagnosed as overtrained, but release returned to levels of the control group after the 4-week period of recovery. All other hormonal responses were variable, but not significantly different. The authors interpreted the suppressed hormonal release (growth hormone, ACTH, cortisol) as being due to a glandular "exhaustion" brought on by the chronic stress of excessive, hard training. Several other studies report related findings for the hormone testosterone in overtrained men (27, 41, 64).

Boyden et al. (10) examined women runners who had increased their weekly training volume (kilometers run per week [km•wk⁻¹]) over a one-year period, and in so doing allowed themselves fewer days of rest per week. These runners received bolus gonadotropin-releasing hormone (GnRH) injections before their training mileage was altered, after a 48 km•wk⁻¹ increase, and again after an 80 km•wk⁻¹ increase. Pituitary responsiveness for the release of luteinizing hormone (LH) and follicle-stimulating hormone to the GnRH stimulus became less and less as the training load increased. Hackney et al. (44) examined male distance runners who where performing 100–200 km•wk⁻¹ in training to look at the pituitary response to a GnRH injection challenge. In this study they compared these trained men to a group of age-matched sedentary men for the amount of LH released after GnRH injection. The trained men were found to have a suppressed response for LH (only ~50% of that in the sedentary men). These LH findings in response to GnRH injection are similar to those reported by MacConnie et al. (71) in highly trained male marathon runners. The reasons for the altered pituitary responses to GnRH injections in each of the studies above were unclear to the authors. Notwithstanding, it could be suggested that a down-receptor phenomena or similar alterations was the potential causative factor, based upon the results of similar work using animal-models (98, 105). Interestingly, Duclos et al. (20) and Hackney et al. (40) have also reported that endurance trained men have an inappropriate release of LH from the pituitary in response to an exercise bout. This has been interpreted as indicating that these trained men have developed a dysfunction within their hypothalamic-pituitary-gonadal hormonal regulatory axis (3, 38, 39). This conclusion is supported by the findings of greatly lowered resting, basal levels of total testosterone, free-testosterone, LH, and prolactin compared to similar age untrained men (3, 38, 39, 50).

Lehmann and associates conducted a six-week exercise study in men employing progressive-overload training in an attempt to induce overtraining (67). These investigators tested pituitary responsiveness to stimulatory drug injections before training (i.e., baseline), after training, and 3 weeks after a regeneration period at the end of the training period. After training and after the regeneration period, ACTH responses to corticoid releasing hormone injection were elevated above baseline, but subsequent cortisol responses to the ACTH release were lower than baseline levels at each of the times. The authors concluded that a decreased sensitivity within the adrenal cortex to ACTH, mostly likely due to reduced peripheral receptor number, was the causative factor for the findings. This finding of reduced ACTH response was similar to that of Wittert et al. (110) denoted earlier.

5.1. Summary

The hormonal alterations noted in these few studies are not in total agreement and are limited in some respects due to their human-based "in vivo" protocols. Nevertheless, they provide some degree of insight as to the type and severity of hormonal alterations that can occur with chronic, hard exercise training. The specific physiological mechanism and causative factors for the alterations remained to be delineated. Notwithstanding, some general conclusions can be made from the findings. Exercising training has the potential to impact upon the regulatory axis controlling neuro-endocrine glands. The impact appears to occur both at the central hypothalamic-pituitary level as well as the peripheral endocrine gland level and result in altered hormonal levels (12, 105). Evidence at this time points to the alterations as being due to possibly: *a*) a reduced number of glandular receptors responding to the humoral-hormonal-neural stimuli ("signal reception defect") sent to an endocrine gland; *b*) a reduced humoral-hormonal-neural stimuli ("signaling defect") sent to an endocrine gland; c) a suppressed glandular-cell production capacity for hormones ("biochemical synthesis defect"); and *d*) some combination of these three factors acting synergistically (17, 74, 103, 104, 105).

6. CONCLUSIONS

The total amount of exercise research literature dealing with the neuro-endocrine system and recovery is extremely small. However, from what is available there does appear to be some tentative, preliminary conclusions that can be postulated. When the neuro-endocrine system is subjected to the stress of exercise there is an acute reactive response (typically characterized by rapid, transit elevations in circulating hormonal concentrations). If adequate rest and recovery are allowed following this acute reactivity phase, then the hormonal responses adjust and return essentially to baseline levels relatively rapidly. The length of time of this phase appears to be a few minutes to a few hours in duration. However, if the acute reactivity response is followed with further increases in training stress too soon, or without some degree of rest-recovery between sessions, there is the onset of a more chronic reactivity response. This is characterized by more prolonged elevations in circulating hormonal levels. The length of this phase appears to be a few hours to a few days. Relative to exercise training this may be coinciding with the point where overload and over-reaching training are occurring in athletes (64). Short periods (i.e., just a few days) of exposure to this stress seem to be tolerated well, but extended period of exposure to this stress are counter-productive. However, if the training stress is pushed still further then a more pronounced chronic reactivity response that is characterized by suppressed circulating hormonal develops. This type of response may result in the development of some degree of aberrant, mal-adaptation at the neuro-endocrine gland, which may manifest itself in the form of a regulatory axis disruption. This late chronic phase response may involve neuro-endocrine alterations that can last for a long period of time (possibly weeks or months) unless an adequate period of regeneration is allowed to occur. This last chronic phase response most likely represents a point where exercise training is inducing "negative maladaptation" as opposed to a positive adaptation. Relative to the sportsman, this is the time where the athlete is most likely developing, or has developed, the "over-training syndrome" (41).

This hypothetical explanation is a "preliminary working model" which may serve as a framework from which other investigators can add or subtract to. This is also a very simple approach attempting to explain the nature of the neuro-endocrine system responses to the interaction between the degree of exercise training versus rest-recovery time. It is recognized that in reality, this is a highly complex, multi-factorial area where many facets are coming into play to affect the neuro-endocrine system's responses and adaptations. For example, the time-line for the changes of individual hormones (i.e., temporal relationship) through the proposed acute and chronic reactivity phase of the model is potentially highly variable. A great deal of future research is necessary in the area of exercise endocrinology to allow a more clear and full discernment and understanding of these interactions.

ACKNOWLEDGMENTS

The author wishes to thank Anna Styers, Grace Hackney, and Dr. Manfred Lehmann for their assistance in the preparation and critical evaluation of the manuscript. During the preparation of this work the author was a Fulbright Scholar at the Lithuanian Institute of Physical Culture, Department of Physiology-Biochemistry, Kaunas, Lithuania. The help and support of the faculty, staff, and students of the institute are also acknowledged.

REFERENCES

1. Aakvaag A, Sand T, Opstad PK, Fonnum F (1978) Hormonal changes in serum in young men during prolonged physical strain. Eur J Appl Physiol 39: 283–291
2. Alen A, Parkarinen A, Hakkinen K, Komi P (1988) Responses of serum androgenic-anabolic and catabolic hormones to prolonged strength training. Int J Sport Med 9: 229–233
3. Arce JC, DeSouza MJ (1983) Exercise and male factor infertility. Sports Med 15: 146–169
4. Astrand PO, Rodahl K (1970) Textbook of work physiology. McGraw-Hill Book Co. New York: 112–404
5. Barron J, Noakes TD, Levy W, Smith C (1985) Hypothalamic dysfunction in overtrained athletes. J Clin Endocrinol Metab 60: 803–806
6. Barwich D, Klett G., Eckert W, Weicker H. (1989) Exercise-induced lipolysis in patients with central Cushing's disease. Int J Sports Med 1: 120–126
7. Berchold P, Berger M, Cuppers HJ, Herrmann J, Nieschlag E, Rudorff K, Zimmerman H, Kruskemper HL (1978) Non-glucoregulatory hormones (T4, T3, rT3, TSH and testosterone) during physical exercise in juvenile type diabetics. Horm Metab Res 10: 269–273
8. Bloom SR, Johnson RH, Park DM, Rennie MJ, Sulaman WR (1976) Differences in the metabolic and hormonal response to exercise between racing cyclists and untrained individuals. J Physiol (Lond.) 258:1–18
9. Bonifazi M, Bela E, Carl G, Lodi L, Martelli G, Zhu B, Lupo C (1995) Influence of training on the response of androgen plasma concentration to exercise in swimmers. Eur J Appl Physiol 70: 109–114
10. Boyden TW, Paramenter R, Stanforth P, Rotkis TC, Wilmore J (1984) Impaired gonadotropin response to gonadotrophin-releasing hormone stimulation in endurance trained women. Fertil Steril 41: 359–363

11. Brandenberger G, Follenius M (1975) Influence of timing and intensity of muscle exercise on temporal patterns of plasma cortisol levels. J Clin Endocrinol Metab 40: 845–849

12. Brisson G, Nolle MA, Desharnaris D, Tanka M (1980) A possible submaximal exercise-induced hyothalamo-hypothyseal stress. Horm Metab Res 12: 201–205

13. Brooks GA, Fahey TD, White TP (1996) Exercise physiology: Human bioenergetics and its application. Mayfield Publishing Co. Toronto: 144–172

14. Bunt J (1986) Hormonal alterations due to exercise. Sports Med 3: 331–345

15. Buono MJ, Yeager JE, Hodgdon J (1986) Plasma adrenocorticotropin and cortisol responses to brief high-intensity exercise in human. J Appl Physiol 61: 1337–1339

16. Costill DL, Thomas R, Robergs RA, Pascoe D, Lambert C, Barr S, Fink WJ (1991) Adaptations to swimming training: influence of training volume. Med Sci Sport Exerc 23(3): 371–377

17. Cumming DC, Wall SR, Galbraith MA, Belcastro A (1987) Reproductive hormonal responses to resistance exercise. Med Sci Sport Exerc 19: 234–238

18. Dale E, Gerlach D, Whilhite AL (1979) Menstrual dysfunction in distance runners. Obstet Gynecol 54: 47–53

19. Despopoulos A, Silbernagi S (1991) Color Atlas of Physiology. Georg Thieme Verlag Stuttgart: 232–271

20. Duclos M, Corcuff JB, Rashedi M, Fougere B (1996) Does functional alterations of the gonadotropic axis occur in endurance trained athletes during after exercise? A preliminary study. Eur J Applied Physiol 73: 427–433

21. Davies CTM, Few J (1972) Effect of exercise on adrenocortical function. J Appl Physiol 35: 688–691

22. Dessypris A, Kuoppasalmi H, Adlercreutz H (1976) Plasma cortisol testosterone androstenedione and luteinizing hormone (LH) in a non-competitive marathon run. J Steroid Biochem 7: 33–37

23. Farrell PA, Kjaer M, Bach FW, Galbo H (1987) Beta-endorphin and adrenocorticotropin response to supramaximal treadmill exercise in trained and untrained males. Acta Physiol Scand 130: 619–625

24. Fellmann N, Coudert J, Jarrige J, Bedu M.m Denis C, Boucher D, Lacour JR (1985) Effects of endurance training on the androgenic response to exercise in man. Int J Sports Med 6(4): 215–219

25. Fry AC, Kraemer W (1997) Resistance exercise overtraining and overreaching: neuroendocrine responses Sports Med 23: 106–129

26. Fry RW, Morton AR, Garcia-Webb P, Crawford GPM, Keast D (1992) Biological responses to overload training in endurance sports. Eur J Appl Physiol 335–344

27. Fry RW, Morton AR, Keast D (1991) Overtraining in athletes: an update. Sports Med 12: 32–65

28. Fry RW, Morton AR, Garcia-Webb P, Keast D (1991) Monitoring exercise stress by changes in metabolic and hormonal responses over a 24-h period. Eur J Appl Physiol 63: 228–234

29. Galbo H (1983) Hormonal and metabolic adaptation to exercise. Stuttgart Georg Thieme Verlag: 2–117

30. Galbo H, Hummer L, Peterson IB, Christensen NJ, Bie W (1977) Thyroid and testicular hormonal responses to graded and prolonged exercise in men. Eur J Appl Physiol 36: 101–106

31. Galbo H, Kjaer M, Mikines KJ (1989) Neurohormonal system. In: Skinner J, Corbin, Landers D, Martin P, Wells CL (eds) Future directions in exercise ad sport science research. Human Kinetics Publishers Champaign: 39–345

32. Gary AB, Telford RD, Weidemann MJ (1993) Endocrine response to intense interval exercise. Eur J Appl Physiol 66: 366–371

33. Gastmann U, Lehmann M, Fleck J, Jeschke D, Keul J (1993) Influence of a 6 -week controlled training on behavior of catecholamines and catecholamine sensitivity in recreational athletes In: Tittel K, Arndt KH, Hollmann W (eds) Sportsmedizin: gestern-heute-morgen Barth Leipzig: 191–193

34. Gawel MJ, Alaghband-Zadeh J, Park DM, Rose FC (1979) Exercise and hormonal secretion. Postgraduate Med J 55: 373–376

35. Goldfarb A, Hatfield BD, Potts J, Armstrong D (1991) Beta-endorphin time course of response to intensity of exercise: effect of training status. Int J Sports Med 12: 264–268

36. Griffin JE. (1996) The thyroid. In: Griffin JE, Odjeda JE (eds) Textbook of endocrine physiology. 3rd ed. Oxford University Press New York: 260–283

37. Guezennec Y, Leger F, Hostr FL, Aymonud M, Pesquies PC (1986) Hormonal and metabolic responses to weightlifting training sessions. Int J Sport Med 7: 100–105

38. Hackney AC (1989) Endurance training and testosterone levels. Sports Med 8: 117–127

39. Hackney AC (1996) The male reproductive system and endurance exercise. Med Sci Sport Exerc 28: 180–189

40. Hackney AC, Fahrner CL, Stupnicki R (1997) Reproductive hormonal responses to maximal exercise in endurance trained men with low testosterone levels. Exp Clin Endocrinol Diabetes 105: 291–295

41. Hackney AC, Pearman SN, Nowacki JM (1990) Physiological profiles of overtrained and stale athletes: a review. J Appl Sport Psych 2: 21–33

42. Hackney AC, Premo MC, McMurray RG (1995) Influence of aerobic and anaerobic exercise on the relationship between reproductive hormones in men. J Sport Sci 13: 305–311

43. Hackney AC, Ness RJ, Schrieber A (1989) Effects of endurance exercise on nocturnal hormone concentrations in males. Chronobiol 6: 341–346

44. Hackney AC, Sinning WE, Bruot BC (1990) Hypothalamic-pituitary-testicular axis function in endurance-trained males. Int J Sports Med 11: 298–303

45. Hackney AC, Feith S, Pozos R, Seale J (1995) Effects of high altitude and cold exposure on resting thyroid hormone concentrations. Aviat Space Environ Med 66: 325–329

46. Hackney AC, Gulledge TP (1994) Thyroid responses during an 8 hour period following aerobic and anaerobic exercise. Physiol Res 43: 1–5

47. Hackney AC, Hodgdon J (1992) Thyroid hormone changes during military field operations: Effects of cold exposure in the Arctic. Aviat Space Environ Med 63: 606–611

48. Hackney AC, Hodgdon J, Hesslink R, Trygg K (1995) Thyroid hormone responses to military winter exercises in the Arctic Region. Arct Med Res 54: 82–90

49. Hackney AC, Sharp RL, Runyan W, Ness RJ (1989) Relationship of prolactin and testosterone changes in males during intensive training Br J Sport Med 23: 194

50. Hackney AC, Sinning WE, Bruot BC (1988) Comparison of resting reproductive hormonal profiles in endurance trained and untrained men. Med Sci Sports Exerc 20: 60–65

51. Hooper SL, Traeger-MacKinnon L, Gordon RD, Bachmann AW (1993) Hormonal responses of elite swimmers to overtraining. Med Sci Sports Exerc 25: 741–747

52. Houmard J, Costill D, Mitchell J, Park S, Fink W, Burns J (1990) Testosterone, cortisol, and creatine kinase levels in male distance runners during reduced training. Int J Sport Med 11: 41–45

53. Jost J, Weiss M, Weicker H (1989) Unterschiedliche regulation des adrenergeb rezeptorsystems in verschiedenen trainingsphasen von schwimmern und langstreckenläufen In: Böning D, Braumann KM, Busse MW, Maassen N, Schmidt W (Hrsg) Sport Rettng order Risiko für die Gesundheit Deutscher Ärzteverlag Köln: 141–145

54. Kern W, Perras B, Wodick R, Fehm HL, Born J (1995) Hormonal secretion during nighttime sleep indicating stress of daytime exercise. J Appl Physiol 79: 1461–1468

55. Kindermann W, Schmitt W (1985) Verhalten von testosteron im blutserum bei körperarbeit unterschiedlicher dauer und intensität. Deutsch Z Sportmed 36: 99–104

56. Kirwan JP, Costill DL, Flynn MG (1988) Physiological responses to successive days of intense training in competitive swimmers. Med Sci Sport Exerc 20: 255–259

57. Kjäer L, Secher NH, Bach FW, Sheikh S, Galbo H (1989) Hormonal and metabolic responses to exercise in humans: effects of sensory nervous blockage. Am J Physiol 257: 95–100

58. Kjaer M, Secher NH, Bach FW, Galbo H, Reeves DR, Mitchell JH (1991) Hormonal, metabolic and cardiovascular responses to static exercise in man: influence of epidural anesthesia. Am J Physiol 261: E214-E220

59. Kraemer RR, Heleniak RJ, Tryniecki L, Kraemer G, Okazaki N, Castracane V (1995) Follicular and luteal phase hormonal responses to low-volume resistive exercise. Med Sci Sports Exerc 27: 809–817

60. Kraemer WJ (1988) Endocrine response to resistance exercise. Med Sci Sports Exerc 20: S152-S157

61. Kraemer WJ, Patton JF, Knuttgen HG, Marchitelli LJ, Cruthirds C, Damokosh A, Harman E, Frykman P, Dziados JE (1989) Hypothalamic-pituitary-adrenal responses to short-duration high-intensity cycle exercise. J Appl Physiol 66: 161–166

62. Kuoppasalmi K, Naveri N, Harkonen M, Adlercreutz H (1980) Plasma cortisol, androstenedione, testosterone and LH in running exercise of different intensities. Scand J Clin Lab Invest 40: 403–409

63. Kuoppasalmi K, Naveri N, Kosunen K, Harkonen M, Adlercreutz H (1981) Plasma steroid levels in muscular exercise. In: Poortmans J, Niset G (eds) Biochemistry of Exercise IV-B University Park Press Baltimore: 149–160.

64. Kuipers H, Keizer HA (1988) Overtraining in elite athletes. Sports Med 6: 79–92

65. Lehmann M, Foster C, Keul J (1993) Overtraining in endurance athletes: a brief review. Med Sci Sports Exerc 25: 854–862

66. Lehmann M, Gastmann U, Steinacker J, Heinz JN, Brouns F (1995) Overtraining in endurance sports: a short review. Med Sport Boh Slov 4: 1–6

67. Lehmann M, Knizia K, Gastmann U, Petersen G, Khalaf A, Bauer S, Kerp L, Keul J (1993) Influence of 6-week 5 days per week training on pituitary function in recreational athletes. Br J Sports Med 27: 186–192

68. Lehmann M, Lormes W, Opitz-Gress A, Steinacker J, Netzer N, Foster C, Gastmann U (1997) Training and overtraining: an overview and experimental results in endurance sport. J Sports Med Phys Fit 37: 7–17

69. Lehmann M, Schnee W, Scheu R, Stockhausen W, Bachl B (1992) Decreased nocturnal catecholamine excretion: parameter for an overtraining syndrome in athletes? Int J Sports Med 13: 236–242

70. Lukaszewska J, Biczoswa B, Hobilewicz D, Wilk M, Obuchowicz-Fibelus B (1976) Effect of physical exercise on plasma cortisol and growth hormone levels in young weight lifters. Endokrynol Pol 2: 149–158

71. MacConnie S, Barkan A, Lampman RM, Schork M, Beitins IZ (1986) Decreased hypothalamic gonadotrophin-releasing hormone secretion in male marathon runners. N Engl J Med 315: 411–416

72. McMurray RG, Eubanks TK, Hackney AC (1995) Nocturnal hormonal responses to resistance exercise. Eur J Appl Physiol 72: 121–126

73. McMurray RG, Forsythe WA, Mar MH, Hardy CJ (1987) Exercise intensity-related responses of beta-endorphin and catecholamines. Med Sci Sports Exerc 19: 570–574

74. Melchionda A, Clarkson P, Denko C, Freedson P, Graves J, Katch F (1984) The effect of local isometric exercise on serum levels of beta-endorphin/beta-lipotropin. Phys Sportsmed 12: 102–109

75. Morgan WP, Brown DR, Raglin JS, O'Connor PJ, Ellickson KA (1987) Psychological monitoring of overtraining and staleness. Br J Sports Med 21: 107–114

76. Morville R, Pesquies PC, Guezennec CY, Serrurier BD, Guignard M (1879) Plasma variations in testicular and adrenal androgens during prolonged physical exercise in man. Ann Endocrinol (Paris) 40: 501–510

77. Nazar K, Jezova D, Kowalik-Borowka E (1989) Plasma vasopressin, growth hormone and ACTH responses to static handgrip in healthy subjects. Eur J Appl Physiol 58: 400–404

78. Nicklas BJ, Ryan AJ, Treuth MM, Harman SH, Blackman MR, Hurley BF, Rogers MA (1995) Testosterone, growth hormone, and IGF-I responses to acute and chronic resistive exercise in men aged 55–70 yr. Int J Sports Med 16: 445–450

79. Opstad PK, Aakvaag A (1983) The effect of sleep deprivation on the plasma levels of hormones during prolonged physical strain and calorie deficiency. Eur J Appl Physiol 51: 97–107

80. Pestell RG, Hurley DM, Vandongen R (1989) Biochemical and hormonal changes during a 100 km ultramarathon. Clin Exp Pharmacol Physiol 16: 353–361

81. Rahkila P, Hakala E, Alen M, Salminen K, Laatikainen T (1988) Beta-endorphin and corticotropin release is dependent on a threshold intensity of running exercise in male endurance athletes. Life Sci 43: 551–557

82. Remes K, Kuoppasalmi K, Adlercreutz H (1985) Effect of physical exercise and sleep deprivation on plasma androgen levels: modifying effect of physical fitness. Int J Sports Med 6: 131–135

83. Schwarz L, Kindermann W (1989) Beta-endorphin, catecholamines and cortisol during exhaustive endurance exercise. Int J Sports Med 10: 324–328

84. Shephard RJ, Sidney KH (1975) Effects of exercise on plasma growth hormone and cortisol levels in human subjects. In: Wilmore JH (ed) Exercise and Sport Sciences Reviews. Academic Press New York. 3: 1–31

85. Shepley B, MacDougall D, Cipriano N, Sutton JR, Tarnopolsky MA, Coates G (1992) Physiological effects of tapering in highly trained athletes. J Appl Physiol 72: 706–711

86. Stone MH, Byrd R, Johnson C (1984) Observations on serum androgen responses to short term resistance exercise in middle-aged sedentary males. Nat Str Cond Assoc J 5:40–65

87. Stone M, Keith R, Kearney J, Fleck S, Wilson G, Triplett N (1991) Overtraining: a review of the signs symptoms and possible causes. J Appl Sport Sci Res 5: 35–50

88. Stray-Gundersen J, Videman T, Snell P (1986) Changes in selected objective parameters during overtraining. Med Sci Sports Exerc 18 (Suppl): 54

89. Storzo GA (1988) Opioids and exercise, and update. Sports Med 7: 109–184

90. Sutton JR, Coleman MJ, Casey J (1973) Androgen responses to physical exercise. Br Med J 1: 520–522

91. Terjung RL, Tipton CM (1971) Plasma thyroxine and TSH levels during submaximal exercise in humans. Am J Physiol 220: 1840–1845

92. Theintz GE (1994) Endocrine adaptation to intensive physical training during growth. Clin Endocrinol 41: 267–272

93. Tharp GD (1975) The role of glucocorticoids in exercise. Med Sci Sports 7: 6–11

94. Urhausen A, Gabriel H, Kindermann W (1995) Blood hormones as markers of training stress and overtraining. Sports Med 20: 251–276

95. Urhausen A, Kindermann W (1992) Biochemical monitoring of training. Clin J Sports Med 2: 52–61

96. Urhausen A, Kindermann W (1994) Monitoring of training by determination of hormone concentration in the blood - review and perspectives. In: Liesen H, Weiß M, Baum M (eds) Regulations - und Repairmechanismen: Köln; Deutscher Ärzte-Verlag: 551–554

97. Urhausen A, Kullmer T, Kindermann W (1987) A 7-week follow-up study of the behavior of testosterone and cortisol during the competitive period in rowers. Eur J Appl Physiol 56: 528–533

98. VanHouten M, Posner BI, Walsh RJ (1980) Radio-autographic identification of lactogen binding sites in rat medium eminence using 125I-human growth hormone: evidence for a prolactin "short-loop" feedback site. Exp Brain Res 38: 455–459

 99. Viru A (1992) Hormonal and metabolic foundations of training effects: sex differences. Medicina Dello Sport 45: 29–38
100. Viru A (1992) Plasma hormones and physical exercise. Int J Sports Med 13: 201–209
101. Viru A, Karelson K, Smirnova T (1992) Stability and variability in hormonal response to prolonged exercise. Int J Sports Med 13: 230–235
102. Viru A, Smirnova T (1982) Independence of physical working capacity from increased glucocorticoid level during short term exercise. Int J Sports Med 3: 80–83
103. Viru A (1985) Hormones in muscular activity: volume I - hormonal ensemble in exercise. CRC Press Boca Raton: 7–88
104. Viru A, Viru M (1997) Organism's adaptivity in sports training. Medicina Sportiva 1: 45–50
105. Widmaier EP (1992) Metabolic feedback in mammalian endocrine systems. Horm Metab Res 24: 147–153
106. Wilkerson JE, Horvath SM (1980) Plasma testosterone during treadmill running. J Appl Physiol 49: 249–253
107. Winder WW, Hagberg JM, Hickson RR, Ehsani AA, McLane JA (1978) Time course of sympathoadrenergic adaptation to endurance exercise training in man. J Appl Physiol 45: 370–374
108. Winder WW, Hickson RC, Hagberg JM, Eshani AA, McLane JA (1979) Training-induced changes in hormonal and metabolic responses to submaximal exercise. J Appl Physiol 46: 766–771
109. Wirth A, Holm G, Lindstedt G, Lundberg PA, Bjorntorp P (1981) Thyroid hormones and lipolysis in physically trained rats. Metab 30: 237–241
110. Wittert GA, Livesey J, Espiner E, Donald R (1996) Adaptation of the hypothalamopituitary adrenal axis to chronic exercise stress in humans. Med Sci Sport Exerc 28: 1015–1019

15

OVERTRAINING AND THE CENTRAL NERVOUS SYSTEM

The Missing Link?

Romain Meeusen

Vrije Universiteit Brussel
Faculty of Physical Education and Physiotherapy
Pleinlaan 2
B-1050 Brussels, Belgium

1. INTRODUCTION

From a physiological standpoint, exercise can impose a significant amount of stress on an organism. Muscular activity requires coordinated integration of many physiological and biochemical systems. Such integration is possible only if the body's various tissues and systems can communicate with each other. The nervous system is responsible for most of this communication through central command and peripheral adjustments. Regular exercise or training will result in better performance, however this 'challenge of homeostasis' can lead to an disturbed balance between training and recovery. This stress is counteracted by several adaptive and regulation mechanisms. The physiological responses to any disturbance of the body's equilibrium and its 'feed-forward' (60, 61) and 'feed-back' from and to the brain is primarily the responsibility of the endocrine system (59). The endocrine and nervous system work in concert to initiate and control movement and all physiological processes it involves. When all facets of the central nervous and neuroendocrine system are performing in harmony, the ability to coordinate and regulate key physiological and metabolic functions, under the perturbations imposed by physical exercise, is quite remarkable. To date, relatively little attention has been placed on the role of the central nervous system in overtraining and fatigue during exercise and training. This chapter will focus on the possible involvement of the central nervous system in the onset of fatigue during exercise and the role of neurotransmitters and neuromodulators in possible mechanisms that underlie overtraining.

Overload, Performance Incompetence, and Regeneration in Sport, edited by Lehmann *et al.*
Kluwer Academic / Plenum Publishers, New York, 1999.

2. STRESS, ...ABOUT ITS DEFINITION

For many years, the study of stress was generally synonymous with the study of Selye's model of stress. The essential element in Selye's definition of stress is that there is a non-specific effect of any demand upon the body. In the first paper on stress (a letter to *Nature* published on July 4th 1936) Selye defined a syndrome produced by 'diverse nocious agents' (84). He described a non-specific response to non-specific agents as follows:

> Experiments on rats show that if the organism is severely damaged by acute non specific nocious agents such as exposure to cold, surgical injury, spinal shock, **muscular exercise** and intoxicants, a typical syndrome appears, the symptoms of which are independent of the nature of the damaging agent.

This 'general adaptation' syndrome develops in three stages: first there is a general 'alarm reaction', which is followed by a 'resistance phase' in which the 'organs' return to normal or in which they adapt. Finally, if the insult continues at the same high level, resistance is lost and the 'exhaustion phase' appears. More or less pronounced forms of these three-stage reaction represents the usual response of the organism to stimuli as cold, exercise, training and overtraining (Figure1).

The first theories about stress were built around the concept that a 'disturbance of homeostasis' will cause a 'stress response'. When the body or organs are confronted with stimuli that disturb the 'intern equilibrium', they will start a 'counter reaction' to correct the disturbance and to reinstate the original homeostasis. This more medical or physiological concept was adapted over the years to a more general model of disturbance, including 'emotional' or 'mental stress'. Today, the most frequently reported stress stimuli in literature are so-called 'emotional loads'. These emotional loads are probably most commonly cited or used reasons for stress responses, although in exercise physiology, the stimulus

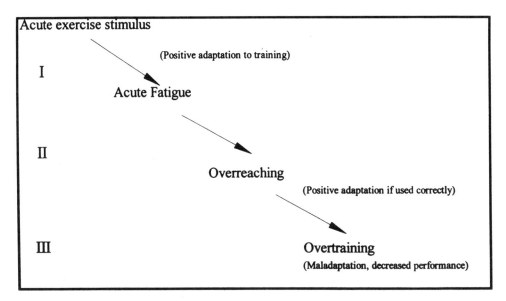

Figure 1. The training–overtraining continuum. (I) Exercise can be considered as a stress stimulus that will create a 'general alarm reaction'. (II) When intensive training is continued or intensified, a resistance phase will appear with a possible positive adaptation (compensation–supercompensation). (III) Finally, maladaptation, with decreased performance are features of the 'exhaustion phase'.

that causes a load to the organism is muscular work, staying very close to Selye's original definition. Since Selye's concept, psychologists redesigned the 'linear model' which is a simple and more physiological explanation, to a transactional model. In this model, the outcome was not simply a function of the eliciting situation, but a joined function of the way the situation is perceived and the efficacy of strategies used by those exposed to it. It describes the way in which the 'thinking body' behaves as a reaction to disturbances of its resting state. These 'coping strategies' are defined as the intervening variables that the subject interposes between the (stress) situation and himself (27). In physiology-oriented research, the relevant concept is that of arousal or activation, again a more linear approach, in which the organism reacts to a potential (or real) threat, with a general, nonspecific response. This 'stress response' is an integral part of an adaptive biological system. Both behavioural and physiological responses to stress are required from humans and other animals to function within the confines of a dynamic and frequently challenged environment (94), in this case the challenge of exercise. When exercise becomes overwhelming, or when the body is trained beyond its ability to adapt (third stage of the general adaptive mechanism of Selye's theory), it becomes overloaded or 'overtrained'. At that time exercise becomes a distress.

3. CENTRAL FATIGUE (A HYPOTHESIS)

Decrements in prolonged exercise performance are traditionally attributed to peripheral (i.e. muscular) factors. This peripheral fatigue has been well studied and can involve impairments at the neuromuscular transmission, at the cellular level, involve substrate depletion, or accumulation of metabolites (70, 31, 45, 39, 29, 24, 30). Recently, there has been some interest in the possible role of the central nervous system in fatigue. A few mechanisms have been suggested to cause central fatigue, such as accumulation of ammonia in the brain (9), and changes in neurotransmitter concentrations especially changes in brain 5-hydroxytryptamine (5-HT serotonin) concentration and metabolism (75). The possibility of a centrally mediated fatigue during exercise was already discussed by Romanowski & Gabriec (81) in the mid-1970s. They linked serotonin to a possible inhibition of brain oxidoreductive processes, while others pointed out the role of dopamine (DA) (22, 15, 12, 44, 48, 49), acetylcholine (Ach) (26), glutamine (67), or γ-aminobutyric acid (GABA) (1) in the onset of central fatigue. Most of the research data deal with the possible role of 5-HT as an attractive candidate for the 'central fatigue hypothesis'. This is probably due to the fact that 5-HT has been shown to induce sleep, affect tiredness, depress motor neuron excitability, influence autonomic and endocrine function, suppress appetite, and is involved in mood and depression (3, 10, 63, 75, 31). It is supposed that increased concentrations of brain 5-HT, will be the key factor in producing central fatigue (75, 11). Increases in brain 5-HT synthesis occurs in response to an increase in the delivery to the brain of tryptophan (TRP). The variability of neurotransmitter release is regulated by a number of processes. One of these presynaptic processes which modulates neurotransmission is the change in neurotransmitter synthesis resulting from the metabolic consequences of eating or exercise (99).

Since the biosynthesis of serotonin is tightly controlled by the activity of its rate-limiting enzyme tryptophan hydroxylase, increases or decreases of its substrate, tryptophan, will trigger increases or decreases in 5-HT synthesis and metabolism (16). TRP and the large neutral amino acids (LNAA), including the branched chain amino acids (BCAA: valine, leucine, isoleucine) use the same carrier to enter the brain, and therefore are competitors for transport over the blood brain barrier. The blood concentration of free TRP or the ratio of free TRP/other LNAA are important parameters for this competition (41, 42, 40). Concentrations of circulating total and free TRP in plasma, and the ratio free TRP/other LNAA depend on several factors i.e. the rate of lipolysis, the activity of hepatic TRP pyrrolase, the uptake into the peripheral and central tissues (16). As free fatty acid (FFA) levels increase during endurance exercise, the amount of TRP bound to albumin is reduced, increasing the level of free TRP in the blood. Other factors such as a high carbohydrate meal (41), insulin administration (41), administration of L-TRP (85), or a combination of these factors will increase the concentration of free TRP in plasma (47). Since brain 5-HT synthesis depends on the plasma concentration of TRP, treatments that elevate plasma TRP will promote accelerated 5-HT synthesis and/or metabolism (15). In this context the question can be asked if it is free tryptophan or total (free plus albumin bound) tryptophan that controls brain tryptophan levels. This question is still a matter of debate, and it is pertinent to answer that the identity of the substrate depends on the experimental conditions (19). For example, stress-induced elevation in brain tryptophan levels following immobilisation is due to an increased entry of total tryptophan, whereas starvation-induced increases in brain tryptophan depend on lipolysis, and in turn, increased availability of free tryptophan for its transporter (16, 19). Since the introduction of the microdialysis technique to collect extracellular fluid, the precursor-induced release of 5-HT has been studied. Some studies (13, 14, 96) found increases, while others (78, 86, 88) did not find any increase in 5-HT release following L-TRP administration. However it is noteworthy that in these studies TRP was administered to control undisturbed animals, while there are several lines of evidence to suggest that the effect of L-TRP on 5-HT function depends on the level of serotonergic neuronal activity (85, 52, 16, 53). For example, after TRP load extracellular 5-HT release will be more effectively in food deprived animals (82, 69), and L-TRP load in combination with electrical stimulation of the raphe nucleus, will increase 5-HT release in the hippocampus dose dependently to the stimulus frequency (85). It was also shown that L-TRP administration, acute exercise or the combination of both treatments, as well as restraint stress were able to increase hippocampal 5-HT release (69, 91) (Figures 2 and 3).

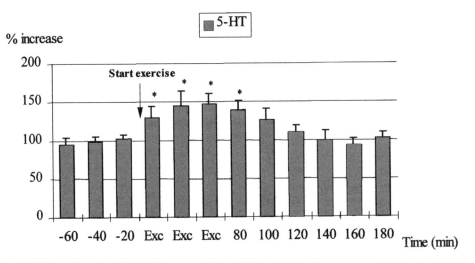

Figure 2. Effects of exercise (12 m/min) on extracellular 5-HT levels in the hippocampus of the food-deprived rat. Microdialysis technique was used to collect 5-HT from the extracellular space before, during and following exercise. Extracellular 5-HT levels increased significantly during 60 min of running and in the first post-exercise sample. Although 5-HT release increased significantly, there was no influence on running performance.

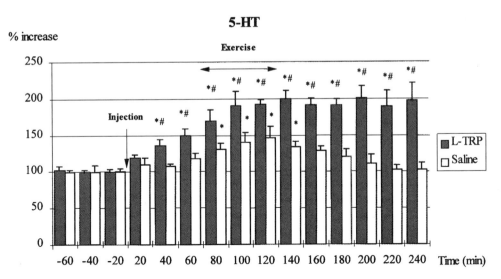

Figure 3. Effects of combined tryptophan (50 mg/kg) and exercise (60min at 12 m/min) on extracellular 5-HT levels in the hippocampus of the food-deprived rat. Microdialysis technique was used to collect 5-HT from the extracellular space before, during and following exercise. Extracellular 5-HT levels increased significantly following TRP injection (no significant increase in saline treated controls). Exercise (from min 60 to 120) significantly increased 5-HT release in both groups. The higher increase of 5-HT release in the tryptophan treated group did not influence running performance. At the end of the experiment (4 hours after injection) extracellular 5-HT levels returned to baseline in the 'saline-exercise' group, while they were still significantly higher in the 'tryptophan-exercise' group. The latter animals did not show any sign of the so-called 5-HT syndrome. *significantly different from baseline; #significant difference between tryptophan and saline (Meeusen et al. 1996).

In our microdialysis study (69), we examined whether exercise-elicited increases in brain tryptophan availability (and in turn 5-HT synthesis) alters 5-HT release in the hippocampus of food-deprived rats. To this end, we compared the respective effects of acute exercise, administration of tryptophan, and the combination of both treatments, upon extracellular 5-HT and 5-hydroxyindole-acetic acid (5-HIAA) levels. All rats were trained to run on a treadmill before implantation of the microdialysis probe and 24 h of food deprivation. Acute exercise (12 m/min for 1 h) increased in a time-dependent manner extracellular 5-HT levels, these levels returning to their baseline levels within the first hour of the recovery period (figure 2). Acute administration of a tryptophan dose (50 mg/kg i.p.) that increased extracellular 5-HIAA (but not 5-HT) levels in fed rats, increased within 60 min extracellular 5-HT levels in food-deprived rats. Whereas 5-HT levels returned toward their baseline levels within the 160 min that followed tryptophan administration, extracellular 5-HIAA levels rose throughout the experiment. Lastly, treatment with tryptophan (60 min beforehand) before acute exercise led to marked increases in extracellular 5-HT (and 5-HIAA levels) throughout the 240 min that followed tryptophan administration (figure 3).

This study indicates that exercise stimulates 5-HT release in the hippocampus of fasted rats, and that a pretreatment with tryptophan (at a dose increasing extracellular 5-HT levels) amplifies exercise-induced 5-HT release. It should be noted that in this study none of the animals showed any sign of fatigue during the exercise session, although extracellular 5-HT levels increased markedly, especially in the L-TRP and exercise trial. Furthermore, since it was shown that during exercise 5-HT, DA, NA and GLU release increases in striatum (72, 70), as well as 5-HT release in hippocampus (69) without affecting running capacity of the animals, the direct relationship between increased 5-HT release and fatigue could not be established. This indicates that much more research is necessary to discover the possible relationship between 5-HT, exercise and fatigue. Changes in neurotransmission caused by eating, precursor loading, and by exercise can thus affect all of the behavioural and physiological functions that precursor dependent neurons happen to subserve (70). This substrate-dependent increase in brain neurotransmitter concentration (and neurotransmission under stimulated conditions) has been the key point in the 'central fatigue hypothesis', which provides support for an inhibitory effect of central serotonergic systems on endurance. Several researchers have begun to look into the validity of the hypothesis that nutritional strategies (BCAA and/or carbohydrate supplementation) could delay fatigue in animals as well as in humans. We will not go into details of all these studies which were reviewed previously (70, 31, 62), but taken together these studies indicate that fatigue can be delayed by ingesting carbohydrates, which are an important energy source for muscle and brain function (25). They further point out that until today there is little evidence to support the hypothesis that BCAA supplementation will increase performance (70, 31), especially since many questions remain regarding the hypothesised mechanisms of action (63). Furthermore, it should not be assumed that a treatment associated with prevention of exercise-elicited changes in serotonergic systems on the one hand, and reduction of fatigue on the other, represents a causal relationship (19). Although the 'BCAA-5-HT-Fatigue' relationship may prove to be correct in the future, the possibility that the positive effects of BCAA's are due to other systems cannot be ignored (19).

4. CENTRAL FATIGUE (OTHER NEUROTRANSMITTERS AND INTERACTIONS)

The lack of reasonable hypothesis to explain central fatigue and central mechanisms during training and overtraining in humans is probably due to the fact that these hypothesis are based on explaining mechanisms that occur in the brain, while until now it is practically (and ethically) impossible to perform real-time measurements in human brains. Therefore, we need to rely on animal studies, or on human experiments in which effects of 'central manipulations' can be measured in peripheral tissues. However, if we want to learn about the central nervous system mechanisms underlying exercise, training, and possibly overtraining, we need to go deeper into neurotransmitter actions (and interactions) during exercise. The role of serotonin in the regulation of motor mechanisms is complex. Thus, a depletion of brain and spinal 5-HT, as well as an increase in the availability of central serotonin, can result in a decrease or an increase in motor activity depending on the experimental model used (50). There are numerous levels at which central 5-HT can affect motor behaviour, from sensory perception, sensory-motor integration, to motor effector mechanisms. 5-HT is endowed with a huge variety of functions that are mediated by more than a dozen receptors. 5-HT_{1A} receptors are located predominantly in brain regions concerned with mood and anxiety (e.g. the limbic system). They are located either at the postsynaptic or the presynaptic level (in the raphe nuclei in the midbrain where they function as autoreceptors that negatively control the tone of serotonergic neurons). In both cases, 5-HT_{1A} receptors mediate neuronal inhibition (19). Several studies (50, 17) investigated the effects of systemically administered 8-OH-DPAT (a 5-HT agonist with preference for the 5-HT_{1A} binding site) on motor activity in open-field locomotion, and on treadmill running in rats. These studies showed that spontaneous locomotor activity was dose-dependently decreased. It should be mentioned that the administration of 8-OH-DPAT and other drugs including precursors, agonists, and releasers can produce various signs of the so-called 'serotonin syndrome' (70). This syndrome is characterised by hyper-activity, head shakes or 'wet dog' shakes, hyper-reactivity, tremor, rigidity, hindlimb ab-duction, straub tail, lateral head weaving, and reciprocal forepaw treading (50). These behavioural signs are sometimes used as an indication for central 5-HT activity (50, 17). It was found that at a low dose 8-OH-DPAT increased locomotion, and that at the higher doses locomotion was inhibited (97). When other 5-HT receptor agonists and antagonists were tested, it was shown that excessive running in the semi-starved rat is suppressed by activation of the $5-HT_{2C}$ receptors and that activation of presynaptic $5-HT_{1B}$ receptors re-sulted in a decreased 5-HT release (97).

These results were confirmed in several studies (6, 57, 66). When animals were treated with mCPP ($5-HT_{2C}$ agonist), run time to exhaustion decreased in a dose-response manner (6), or a clear-cut hypolocomotion was induced by mCPP (57, 66). Bailey et al. (7, 8) examined the effects of quipazine dimaleate (QD), a general 5-HT agonist with high af-finity for the $5-HT_3$ receptor (97), and LY53857, a 5-HT antagonist specific to $5-HT_{2C}$ and $5-HT_{2A}$ receptors (and catecholamine receptor blocker activity at a high dosage). Run time to exhaustion was reduced in a dose-dependent manner by increasing dosages of QD, while the time to exhaustion was increased with LY53857 administration but only at the highest dose (7, 8). The results also indicated a possible importance of the interaction be-tween brain 5-HT and DA in the onset of fatigue. The role of DA, 5-HT, and/or the inter-action with other neurotransmitters was investigated by several authors who used precursor loading, and/or pharmacological manipulation on exercise performance in ani-mals and humans. It has been shown that pretreatment of exercising rats with ampheta-mine (AMPH), which releases DA, or apomorfine (48) (a DA agonist), extends the time to exhaustion in fatigued and non-fatigued rats at low doses, while at high doses it reduces time to exhaustion (6, 20). The results of several studies (48, 49) suggest that the time to exhaustion is influenced by the activity of nigrostriatal dopaminergic neurons (48). Cen-tral DA depletion could hasten time to exhaustion, while increasing central dopaminergic activity prolongs time to exhaustion (48). During the later phase of the run to exhaustion there is probably a need to increase striatal dopaminergic activity. If this is the case, dopaminergic agonists may improve exercise capacity by facilitating such recruitment (49). On the other hand, these authors state that dopamine accumulation in the striatum late in exercise as found in their studies (48, 49), may reflect decreases in dopamine re-lease, perhaps due to activation of dopamine autoreceptors (49). We have found that AMPH given 20 min before, or at the onset of exercise (60 min run), significantly in-creased DA release in striatum compared to saline treated controls (Figure 4). In this ex-periment running performance was not influenced, but at the end of the experiment (more than 2 h post exercise), striatal extracellular DA levels in the AMPH group were signifi-cantly lower than in controls, possibly indicating an autoreceptor mediated phenomenon (Meeusen et al., unpublished data). However, effects of cortical noradrenergic influence through α_1-adrenergic receptors should not be ruled out, since stimulation of these recep-tors at the cortical level increases the release of subcortical DA (28). This again empha-sises the importance of neurotransmitter interactions (Figure 4).

Two studies (32, 98) examined the effect of a 5-HT reuptake inhibitor on time to ex-haustion in humans. Both found a decrease in exercise time to exhaustion with the reup-take inhibitor compared to placebo. However, very recently we performed a double bind cross-over study in which the effect of a single dose of a 5-HT reuptake inhibitor (Fluoxet-

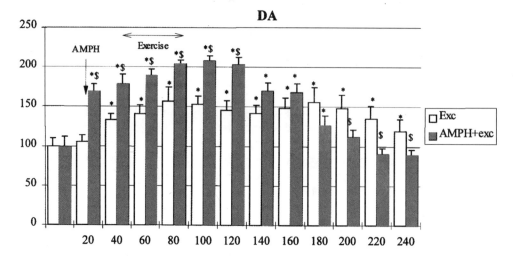

Figure 4. Effects of 0.5mg/kg Amphetamine (AMPH) or saline followed by 60 min of exercise on striatal dopamine (DA) release. Injection of 0.5mg/kg AMPH significantly increases DA release in the first dialysate (70% increase). Exercise significantly increases DA release in both groups during and 2 hours following exercise. Extracellular DA levels are significantly higher in the AMPH treated group during and 40 min following exercise. At the end of the experiment DA release is lower compared to saline treated animals. * significant increase compared to baseline. $ significant difference between saline and AMPH group.

ine-HCL) on exercise performance during a 90 min time trial at 65% $Watt_{max}$ was examined. The measure of performance was the time to complete the target amount of work. The results showed no difference in performance between experimental and placebo group (Meeusen et al. 1998 manuscript in preparation). These results are in line with our previous studies (33, 34, 35, 71, 74). De Meirleir (33) examined the influence of a DA agonist (pergolide) and a serotonin antagonist (ketanserin) on exercise performance. The results showed that oral treatment with the serotonin (5-HT$_{1C}$/5-HT$_2$) antagonist had no influence on exercise performance. It did not alter heart rate at rest or during exercise, but it elicited a shift to the right of the lactic acid curve (33). The DA agonist (D$_1$/D$_2$) lowered heart rate, systolic blood pressure, and enhanced maximal work capacity (33, 34).

Since literature indicates that exercise performance can be decreased with DA antagonism and 5-HT agonism, we examined if the opposite was also true, i.e. the effect of the administration of a DA precursor (L-DOPA) or a specific 5-HT$_{2A/2C}$ antagonist (Ritanserin) on time to exhaustion in humans was examined. The results showed that neither a metabolic precursor of DA nor a specific centrally acting 5-HT$_{2A/2C}$ antagonist influenced the time to exhaustion on a bicycle trial at 65% $Watt_{max}$ (74). The results from these studies emphasise the importance of DA and 5-HT (and probably other neurotransmitter) interactions during exercise. Central 5-HT$_3$ receptors might be involved in the observed behaviour because they interact with the dopaminergic neurotransmitter system (97). LY53857 is a potent and selective 5-HT$_2$ receptor antagonist, but has also affinity for α_2-receptors, therefore at the highest dose can interact with catecholaminergic receptors (18) Taken together, these results indicate a role for several DA and 5-HT transmitter receptors in motor control (and maybe 'central fatigue'). However, any possible role of 5-HT, DA (and other transmitters) in motor function should be perceived as a continuum (5). This continuum is not only important at the brain level, but has its own importance in the interaction between central neurotransmission and the peripheral processes during exercise, including the neuroendocrine system, especially the activity of the hypothalamic-pituitary-adrenal axis (HPA). Neurotransmitter systems not only influence each other, but they also are intimately linked to the HPA (16, 36, 38).

5. CENTRAL NERVOUS SYSTEM: OVERTRAINING–NEUROTRANSMISSION

To adjust the disturbance in resting homeostasis produced by an exercise stimulus, a number of regulatory systems are called upon to return the body to a new level of homeostasis. Principal among these is the central nervous system, which is capable of making very rapid adjustments to large segments of the body, and the endocrine system which can have a more global and far reaching effect but requires more time to respond (68). The catecholamines, both as a neurotransmitter and as a hormone, have very powerful regulatory properties that exert control over a number of critical physiological and metabolic functions central to the ability to sustain physical exercise (68). The symptoms associated with overtraining, such as changes in emotional behaviour, sleep disturbances, and hormonal dysfunctions are indicative of changes in the regulation and coordinative function of the hypothalamus (56). The HPA, together with the autonomic nervous system, is the most important stress system in the body. There is evidence that central serotonergic systems act upon the sympathetic nervous system and the HPA and that reciprocally, glucocorticoids and catecholamines (derived from sympathetic nerves and the adrenomedulla) affect central serotonergic systems (16). In pathological situations such as in major depression (37) and probably also in overtraining, the glucocorticoids and the brain monoaminergic systems apparently fail to restrain the HPA response to stress. In this last part we will try to formulate some possible strategies which could underlie the reciprocal influence (disturbance) of the central nervous system and peripheral markers during overtraining. Overtraining is frequently defined as an imbalance in autonomic (sympathetic-parasympathetic) functioning. Fatigue and apathy dominate in the parasympathetic type of overtraining, which is typical for endurance sports.

Restlessness and hyperexcitability dominate in the sympathetic type of overtraining, which is more typical in explosive sports or related to additional significant nontraining stress factors (65). Other effects of strenuous exercise and (over)training such as mood state changes, sleep disturbances, changes in recovery rate, prolonged feelings of fatigue and changes in reproductive state all indicate maladaptation in certain parts in the brain with consequent changes in hypothalamic effector output (56). During overtraining peripheral catecholaminergic drive is disturbed, with possible reduced β-adrenoreceptor density at various target organs (65). The afferent catecholaminergic drive from the working muscles is known to produce a 'feed-forward' stimulation of the neuroendocrine and motor centres in the brain (59). These, and other (e.g. testosteron, cortisol, CRH, ACTH) peripheral markers for overtraining clearly indicate that in a state of overtraining signals from the hypothalamus and the 'higher' centres that control hypothalamic drive will be influenced not only during exercise and training, but might also be disregulated during overtraining. As it has been shown that exercise and training influences neurotransmitter release in various brain nuclei (72, 73, 69, 71), possible disregulation at this level could play a key role in the maladaptation to the 'stress' of exercise, training and overtraining.

Recently we found (71) that endurance training decreases basal neurotransmitter outflow in of DA, NA and GLU the striatum. When an acute exercise stimulus (60 min minutes of exercise) was given to these animals, extracellular DA, NA and GLU levels increased in both control and trained animals. The results indicate that exercise training appears to result in diminished basal activity of striatal neurotransmitters, while maintaining the necessary sensitivity for responses to acute exercise. These observations raise the possibility that there could exist a exercise-induced change in receptor sensitivity. Although in these experiments both acute and chronic exercise did not to influence extracellular GABA levels in a significant manner, one should also consider the possible role of GABA in the regulation of the stress response (overload of the central regulation during overtraining). It is well known that in response to stress the GABAergic system undergoes alterations which depend upon the nature and intensity of the stress. GABA influences the activity of the HPA at several levels: it inhibits the release of CRF from the hypothalamus and possibly affects the secretion of ACTH from the anterior pituitary gland and corticosteroidogenesis in the adrenal cortex. In turn, the GABAergic system is influenced by steroid hormones, peptides, purines, excitatory amino acids, and other neurotransmitters (90). Modifications of GABAergic functions by catecholamines or other transmitters change physiological and behavioural responses to stress (90). There is an extensive amount of anatomical, biochemical and physiological data lending support to an excitatory influence of central serotonergic systems upon the HPA (16). Noradrenaline is also linked to the regulation of the HPA, which is dysregulated not only in overtraining (56) but also in conditions such as in major depression and certain anxiety disorders (19, 37). In these conditions glucocorticoids and brain monoaminergic systems apparently fail to restrain the HPA response to stress. Activation of the brain noradrenergic and serotonergic systems during stress is believed to increase the secretion of ACTH by stimulating the secretion of CRH (79, 92). Central serotonergic activity may either inhibit sympathetic nerve discharge by acting on 5-HT_{1A} receptors (21, 80), or creates an sympathoexcitation by acting on 5-HT_{2A} receptors (4, 21). It was shown that activation of 5-HT_{2A} receptors promotes peripheral adrenaline release and hyperglycaemia (16, 21).

Pharmacological data suggest that central serotonergic activity may exert a positive control over adrenomedullary catecholaminergic systems (through 5-HT_{1A}; 5-HT_{2A}), whereas it respectively excites and inhibits some sympathetic nerves through 5-HT_2 receptors and 5-HT_{1A} receptors (16). Furthermore, it was shown recently that moderate and intensive exercise training significantly reduces the efficacy of a selective 5-HT_{1B} receptor agonist in the substantia nigra. A desensitisation of the 5-HT_{1B} receptors was seen, suggesting a regulation mechanism altering the functional efficacy of these receptors and consequently serotonergic activity (83). These results might further indicate that serotonergic drive could also influence dopaminergic activity since dopaminergic cell bodies are located at the substantia nigra. It has been shown previously that there exists a controlling or modulating role of the serotonin system on the function of nigro-striatal and ventral tegmental-limbic dopamine systems (77). The fact that raphe cell bodies are densely innervated with noradrenergic cell terminals, emphasise that the Locus Coeruleus noradrenaline and the dorsal raphe serotonergic system reciprocally innervate each other (92).

Again we have to point out that interaction of serotonergic, dopaminergic, noradrenergic and other neurotransmitter and neuromodulator systems could (directly or indirectly) influence central and peripheral parameters. Traditionally, the main criterion for a stress response is an increase in the secretion of stress hormones. On this basis, a decline in hormonal secretion when stress is repeated or prolonged is commonly interpreted as indicating stress adaptation (89). A large body of evidence indicates that stressful experiences also alter neurotransmitter metabolism and release in several brain areas (2, 43, 46, 51, 54, 55, 58, 64, 76, 95). Repeated exposure to stress may lead to different responsiveness to subsequent stressful experiences depending on the stressor as well as on the stimuli paired with the stressor, either leading to an unchanged or increased or decreased neurotransmitter and receptor function (2, 43, 46, 51, 54, 55, 64, 76, 95). Behavioural adaptation (release, receptor sensitivity, receptor binding etc.) in higher brain centres will certainly influence hypothalamic output (64). It has been shown that immobilisation stress not only increases hypothalamic monoamine release, but consequently CRH and ACTH secretion (87). Chronic stress and the subsequent chronic peripheral glucocorticoid secretion could play an important role in the desensitisation of higher brain centres response during acute stressors, since it has been shown that in acute (and also chronic) immobilisation the responsiveness of hypothalamic CRH neuron rapidly falls (23). These adaptation mechanisms could be the consequence of changes in neurotransmitter release, depletion of CRH and/or desensitisation of hypothalamic hormonal release to afferent neurotransmitter input (23). In overtraining it is assumed that a 'maladaptation' to chronic exercise (and other) stress occurs. There is evidence that apart from other brain areas, the hippocampus downregulates the HPA in both rats and primates. Recently we have analysed the impact of streptozotocin (STZ)-elicited diabetes (a chronic stress for the body) on hippocampal extracellular 5-HT levels both under basal conditions and during restraint stress. Restraint stress is a procedure known to stimulate 5-HT synthesis/metabolism and release. The chronic stress decreased hippocampal 5-HT metabolism, but did not alter baseline 5-HT release (compared to controls). Restraint stress increased extracellular 5-HT levels in controls, but did not increase 5-HT in the diabetic animals (91). This might indicate that in the chronic stressed animals the central response to an acute stressor (restraint) is impaired. The examples cited previously, illustrate that in several brain nuclei chronic stress will create an adaptation mechanism (autoreceptor mediated, neurotransmitter interactions, or other mechanisms). When an animal is confronted with a novel stressful stimulus, a sensitisation will occur. However, in chronic, very intense stress situations (as in streptozotocin diabetes where other peripheral hormonal mechanisms also play an important role) this sensitisation of hippocampal 5-HT release does not occur. One might speculate that in overtraining (the step beyond coping with stress) a comparable mechanism occurs. If these results from a 'non-exercise' chronic stress situation, can be translated to another chronic stress such as overtraining remains to be elucidated. Interestingly, in a recent published overtraining study in humans (93) overtrained athletes showed a depressed exercise-induced increase of ACTH, indicating a suppression of the HPA. In contrast with our animal study, where plasma corticosterone levels were increased (indicating an aversiveness to stress), in overtrained athletes (93) the exercise-induced cortisol levels tended to be lower. We have to point out that the contrasting findings for repeated stress effects upon brain biogenic amine systems might reflect the differential sensitivity of individual brain regions to different types of stress. This again emphasises the difficulties (and dangers) in comparing results from different research paradigms, different stressors, animal and human studies, different stimuli etc. Determination of hormones is problematic since not only their pulsative and/or circadian rhythms, but also the difference in sampling methods, limit comparison. When several overtraining (or chronic stress) studies are compared, the difference in 'overtraining status' i.e., overreaching, short term overtraining and other terms that are used in literature makes comparison of the studies problematic (if not impossible).

6. CONCLUSION

The dualistic view of the mind and the body being distinct entities has long been abandoned in sports research community. The adoption of a holistic approach has resulted in a vast quantity of research designed to define the precise relationships between the brain and the body. However, understanding brain function takes more than just one series of experiments. At this moment it seems that it will probably take more 'brains' than a human has available to understand only a part of brain functioning. Exercise can be considered as a challenge of homeostasis. This stress is counteracted by several adaptive and regulation mechanisms. The central nervous system in combination with the (neuro)endocrine system plays a very important role in this homeostatic regulation. Although the substrate-dependent increase in brain 5-HT could play an important role in the onset of 'central fatigue', it seems that the exact mechanism(s) underlying this hypothesis still needs to be elucidated. Clearly, more than one neurotransmitter and neuromodulator are involved. The multifactorial aspect of central nervous functioning with different effects of increased neurotransmitter release in different brain regions, the complexity of several receptors and the mutual interaction of various neurotransmitters (having excitatory and inhibitory features) illustrates this. While it seems that there are data to support the hypothesis that BCAA with or without carbohydrate may affect the physiological and psychological responses to endurance exercise, today there is no 'clearcut' answer to whether or not BCAA supplementation helps to prolong time to exhaustion. Hormone responses to exercise are organised and modulated by numerous factors. During exercise, the activity of endocrine system is determined by the intensity and duration of muscular activity on the one hand, and by the adaptation of the organism to muscular activity on the other hand. It has often been speculated that a hormonal mediated central disregulation occurs during the pathogenesis of overtraining, but physiopharmacological influences of brain transmitters on hormones of the HPA and sympathetic nervous system (the 'descending' part of the central nervous system HPA-loop) should also be considered in the genesis of overtraining. The HPA is controlled by complex regulatory mechanisms. Several neurotransmitters might play an important role in the onset of the central dysfunctions that occur in the maladaptation to chronic exercise stress or overtraining. We are aware of the fact that this multifactorial disturbance in the onset of overtraining could give rise to more questions than answers, yet it is quite clear that several neurotransmitters and neuromodulators are involved in the disturbed adaptation to chronic exercise stress we call overtraining. The arguments we proposed in this chapter might illustrate the regrettable limitation of research in this field: most investigations have attempted to concentrate on one or just a few aspects of the cascade of events that occurs during exercise, training, overtraining. Many questions remain to be answered regarding the hypothesised mechanisms of action, and the physiological and psychological effects of endurance exercise performance, training, and overtraining. What remains to be seen is how individual elements of the several neurotransmitters and neuromodulators contribute to this scheme: how modulation of pre- and postsynaptic receptors, coupled with rapid or delayed changes in transmitter synthesis and release, accounts for the behavioural and performance impact of a stress called overtraining. Finally, we need not look at the present research data as being totally contradictory or inconsistent, it probably just means that we are only at the beginning of 'the age of the brain'.

REFERENCES

1. Abdelmalki A, Merino D, Bonneau D, Bigard A, Guezennec Y (1997) Administration of a $GABA_A$ agonist baclofen before running to exhaustion in the rat: effects on performance and some indicators of fatigue. Int Sports Med 18: 75–78
2. Abercrombie E, Keefe K, DiFrischia D, Zigmond M (1989) Differential effect of stress on in vivo dopamine release in striatum, nucleus accumbens, and medial frontal cortex. J Neurochem 52: 1655–1658

3. Acworth I, Nicolass J, Morgan B, Newsholme E (1986) Effect of sustained exercise on concentrations of plasma aromatic and branched chain amino acids and brain amines. Biochem Biophys Res Comm 137: 149–153

4. Bagdy G, Szemeredi K, Kanyicska B, Murphy D (1989) Different serotonin receptors mediate blood pressure, heart rate, plasma catecholamine and prolactin responses to m-chloro-phenylpiperasine in conscious rats. J Pharmacol Exp Ther 250: 72–78

5. Bailey S, Davis J (1994) Response to letter to the editor by F. Chaouloff [letter]. Int J Sports Med 15: 340–341

6. Bailey S, Davis J, Ahlborn E (1992) Effect of increased brain serotonergic activity on endurance performance in the rat. Acta Physiol Scand 145: 75–76

7. Bailey S, Davis J, Ahlborn E (1993a) Serotonergic agonists and antagonists affect endurance performance in the rat. Intern J Sports Med 14: 330–333

8. Bailey S, Davis J, Ahlborn E (1993b) Neuroendocrine and substrate responses to altered brain 5-HT activity during prolonged exercise to fatigue. J Appl Physiol 74: 3006–3012

9. Bannister E, Cameron B (1990) Exercise-induced hyperammonemia: peripheral and central effects. Int J Sports Med 11(Suppl 2): S129-S142

10. Blomstrand E, Celsing F, Newsholme E (1988) Changes in plasma concentrations of aromatic and branced-chain amino acids during sustained exercise in man and their possible role in fatigue. Acta Physiol Scand 133: 115–121

11. Blomstrand E, Perret D, Parry-Billings M, Newsholme E (1989) Effect of sustained exercise on plasma amino acid concentrations and on 5-hydroxytryptamine metabolism in six different brain regions in the rat. Acta Physiol Scand 136: 473–481

12. Burgess M, Davis M, Borg T, Buggy J (1991) Intracranial self-stimulation motivates treadmill running in rats. J Appl Physiol 71: 1593–1597

13. Carboni E, Cadoni C, Tanda G, Di Chiara G (1989) Calcium dependent, Tetrodotoxin-sensitive stimulation of cortical serotonin release after tryptophan load. J Neurochem 53: 976–978

14. Carboni E, Cadoni C, Tanda G, Frau R, Di Chiara G (1991) Serotonin release and metabolism in the frontal cortex: effect of drugs, tryptophan and stress. In: Rollema H, Westerink B and Drijfhout J (eds) Monitoring molecules in neurosciences. The Netherlands: Krips Repro: 235–237

15. Chaouloff F (1989) Physical exercise and brain monoamines: a review. Acta Physiol Scand 137: 1–13

16. Chaouloff F (1993) Physiolopharmacological interactions between stress hormones and central serotonergic systems. Brain Res Rev 18: 1–32

17. Chaouloff F (1994a) Influence of physical exercise on 5-HT$_{1A}$ receptor- and anxiety-related behaviours. Neurosci Let 176: 226–230

18. Chaouloff F (1994b) Serotonin$_{1C, 2}$ receptors and endurance performance [letter]. Int J. Sports Med 15: 339 1994b

19. Chaouloff F (1997) The serotonin hypothesis. In: Morgan WP (ed) Physical activity & mental health. Taylor & Francis, Washington: 179–198

20. Chaouloff F, Kennett GA, Serrurier B, Merino D and Curzon G (1986) Amino acid analysis demonstrates that increased plasma free tryptophan causes the increase of brain tryptophan during exercise in the rat. J Neurochem 46: 1647–1650

21. Chaouloff F, Laude D, Baudrie V (1990) Effects of the 5-HT1C/5-HT2 receptor agonist DOI and -methyl-5-HT on plasma glucose and insulin levels in the rat. Eur J Pharmacol 187: 435–443

22. Chaouloff F, Laude D, Meringo D, Serrurier B, Guezennec Y (1987) Amphetamine and alpha-methyl-p-tyrosine affect the exercise induced imbalance between the availability of tryptophan and synthesis of serotonin in the brain of the rat. Neuropharmacol 268: 1099–1106

23. Cizza G, Kvetnansky R, Tartaglia M, Blackman M, Chrousos G, Gold P (1993) Immobolisation stress rapidly decreases hypothalamic corticotropin-releasing hormone secretion in vitro in the male 344/N fischer rat. Life Sci 53: 233–240

24. Coggan A, Coyle E (1991) Carbohydrate ingestion during prolonged exercise: effects on metabolism and performance. Exerc Sport Sci rev 19: 1–40

25. Coggan A, Williams B (1995) Metabolic adaptations to endurance training: substrate metabolism during exercise. In: Hargreaves M (ed) Exercise metabolism. Human Kinetics Publishers Inc. Champaign USA: 177–210

26. Conlay L, Sabounjian L, Wurtman R (1992) Exercise and neuromodulators: choline and acetylcholine in marathon runners. Int J Sports Med 13 (Suppl 1): S141-S142

27. Dantzer R (1993) Coping with stress. In: Stanford S & Salmon P (eds) Stress, from synapse to syndrome. Academic Press, London: 167–189

28. Darracq L, Blanc G, Glowinski J, Tassin J (1998) Importance of the noradrenaline-dopamine coupling in the locomotor activating effects of D-Amphetamine. J Neurosci 18: 2729–2739

29. Davis M (1995) Central and peripheral factors in fatigue. J Sports Sci 13: S49-S53

30. Davis M (1996) Nutritional influences on central mechanisms of fatigue involving serotonin. In: Maughan & Shirreffs (eds) Biochemistry of exercise IX. Human Kinetics Champaign USA: 445–455

31. Davis M, Bailey S (1997) Possible mechanisms of central nervous system fatigue during exercise. Med Sci Sports Exerc 29: 45–57

32. Davis M, Bailey S, Jackson D, Strasner A, Morehouse S (1993) Effects of a serotonin agonist during prolonged exercise to fatigue in humans. Med Sci Sports Exerc [abstract] 25: S78

33. De Meirleir K (1985) Studies on cardiovascular drugs and (neuro)humoral substances in dynamic exercise [PhD thesis] Brussels: Vrije Univ Brussel

34. De Meirleir K, Gerlo F, Hollmann W, Vanhaelst L (1986) Cardiovascular effects of pergolide mesylate during dynamic exercise. Proceedings of the BPS, 633P

35. Dendale P, De Meirleir K (1993) Pergolide mesylate and physical performance: a brief report. Clin J Sports Med 3: 256–258

36. Dishman R (1994) Biological psychology, exercise, and stress. Quest 46: 28–59

37. Dishman R (1997) The norepinephrine hypothesis. In: Morgan WP (ed) Physical activity & mental health. Taylor & Francis, Washington: 199–212

38. Dunn A, Dishman R (1991) Exercise and the neurobiology of depression. Exerc Sport Sci Rev 19: 41–98

39. Enoka R, Stuart D (1992) Neurobiology of muscle fatigue. J Appl Physiol 72: 1631–1648

40. Fernstrom J (1983) Role of precursor availability in control of monoamine biosynthesis in brain. Physiol rev 63: 484–546

41. Fernstrom J, Wurtman R (1971) Brain serotonin content: physiological dependence on plasma tryptophan levels. Science 173: 149–52

42. Fernstrom J, Wurtman R (1972) Brain serotonin content: physiological regulation by plasma neutral amino acids. Science 178: 414–6

43. Finlay J, Zigmond M, Abercrombie E (1995) Increased dopamine and norepinephrine release in medial prefrontal cortex induced by acute and chronic stress effects of diazepam. Neuroscience 64: 619–628

44. Gerald M (1978) Effect of (+)-amphetamine on the treadmill endurance performance of rats. Neuropharmacol 17: 703–704

45. Green H (1997) Mechanisms of muscle fatigue in intense exercise. J Sports Sci 15: 247–256

46. Gresch P, Sved A, Zigmond M, Finlay J (1994) Stress-induced sensitization of dopamine and norepinephrine efflux in medial prefrontal cortex of the rat. J Neurochem 63, 575–583

47. Hernandez L, Parada M, Baptista T, Schwartz D, West H, Mark G, Hoebel B (1991) Hypothalamic serotonin in treatments for feeding disorders and depression as studied by brain microdialysis. J Clin Psychiatry 52 (Suppl): 32–40

48. Heyes M, Garnett E, Coates G (1985) Central dopaminergic activity influences rats ability to run. Life Sci 36: 671–677

49. Heyes M, Garnett E, Coates G (1988) Nigrostriatal dopaminergic activity is increased during exhaustive exercise stress in rats. Life Sci 42: 1537–1542

50. Hillegaart V, Wadenberg M, Ahlenius S (1989) Effects of 8-OH-DPAT on motor activity in the rat. Pharmacol Biochem Behav 32: 797–800

51. Imperato A, Angelucci L, Casolini P, Zocchi A, Puglisi-Allegra S (1992) Repeated stressful experiences differently affect limbic dopamine release during and following stress. Brain Res 577: 194–199

52. Jacobs B (1991) Serotonin and behaviour: emphasis on motor control. J Clin Psy 52, 12(Suppl): 17–23

53. Jacobs B, Fornal C (1993) 5-HT and motor control: a hypothesis. TINS 16:346–350

54. Jordan S, Kramer G, Zukas P, Petty F (1994) Previous stress increases in vivo biogenic amine response to swim stress. Neurochem Res 19: 1521–1525

55. Keefe K, Stricker E, Zigmond M, Abercrombie E (1990) Environmental stress increases extracellular dopamine in striatum of 6-hydroxydopamine-treated rats: in vivo microdialysis studies. Brain Res 527: 350–353

56. Keizer H (1998) Neuroendocrine aspects of overtraining. In: Kreider, Fry, O'Toole (eds) Overtraining in Sport. Human Kinetics, Champaign Illinois: 145–167

57. Kennett G, Curzon G (1988) Evidence that mCPP may have behavioural effects mediated by central 5-HT_{1C} receptors. Br J Pharmacol 94:137–147

58. Kirby L, Chou-Green J, Davis K, Lucki I (1997) The effects of different stressors on extracellular 5-hydroxytryptamine and 5-hydroxyindoleacetic acid. Brain Res 760: 218–230

59. Kjaer M (1989) Epinephrine and some other hormonal responses to exercise in man: with special reference to physical training. Int J Sports Med 10:2–15

60. Kjaer M, Secher N, Bach F, Galbo H (1987a) Role of motor center activity for hormonal changes and substrate mobilisation in exercising man. Am J Physiol 253: R687-R695
61. Kjaer M, Secher N, Galbo H (1987b) Physical stress and catecholamine release. Balliere's Clin Endocrin Metabol 1: 279–289
62. Kreider R (1998) Central Fatigue hypothesis and overtraining. In: Kreider, Fry, O'Toole (eds) Overtraining in Sport. Human Kinetics, Champaign Illinois: 309–331
63. Kreider R, Miriel V, Bertun E (1993 Amino acid supplementation and exercise performance. Sports Med 16: 190–209
64. Lachuer J, Delton I, Buda M, Tappaz M (1994) The habituation of brainstem catecholaminergic groups to chronic daily restraint stress is stress specific like that of the hypothalamo-pituitary-adrenal axis. Brain Res 638: 196–202
65. Lehmann M, Foster C, Netzer N, Lormes W, Steinacker J, Liu Y, Opitz-Gress A, Gastmann U (1998) Physiological responses to short- and long-term overtraining in endurance athletes. In : Kreider, Fry, O'Toole (eds) Overtraining in Sport. Human Kinetics, Champaign Illinois: 19–46
66. Lucki I, Ward H, Frazer R (1989) Effect of 1-(m-chlorophenyl) piperazine and 1-(trifluoromethylphenyl) piperazine on locomotor activity. J Pharmacol Exp Ther 249: 155–164
67. MacLaren D, Gibson H, Parry-Billings M, Edwards R. (1989) A review of metabolic and physiological factors in fatigue. Exc Sport Sci Rev 17: 29–66
68. Mazzeo R (1991) Catecholamine responses to acute and chronic exercise. Med Sci Sports Exerc 23: 839–845
69. Meeusen R, Chaouloff F, Thorré K, Sarre S, De Meirleir K, Ebinger G, Michotte Y (1996) Effects of tryptophan and/or acute running on extracellular 5-HT and 5-HIAA levels in the hippocampus of food-deprived rats. Brain Res 740: 245–252
70. Meeusen R, De Meirleir K (1995) Exercise and brain neurotransmission. Sports Med 20: 160–188
71. Meeusen R, Smolders I, Sarre S, De Meirleir K, Keizer H, Serneels M, Ebinger G, Michotte Y (1997a) Endurance training effects on striatal neurotransmitter release, an 'in vivo' microdialysis study. Acta Physiol Scand 159: 335–341
72. Meeusen R, Sarre S, Michotte Y, Ebinger G, De Meirleir K (1994) The effects of exercise on neurotransmission in rat striatum, a microdialysis study. In: Louilot A, Durkin T, Spampinato U, Cador M (eds) Monitoring Molecules in neuroscience: 181–182
73. Meeusen R, Smolders I, Sarre S, De Meirleir K, Ebinger G, Michotte Y (1995) The effects of exercise on extracellular glutamate (GLU) and GABA in rat striatum, a microdialysis study. Medicine and Science in Sports and Exercise 27: S215
74. Meeusen R, Roeykens J, Magnus L, Keizer H, De Meirleir K (1997b) Endurance performance in humans: the effect of a dopamine precursor or a specific serotonin antagonist. Int J Sports Med 18: 571–577
75. Newsholme E, Acworth I, Blomstrand E (1987) Amino acids, brain neurotransmitters and a functional link between muscle and brain that is important in sustained exercise. In: Benzi G (ed) Advances in myochemistry. John Libby Eurotext, London: 127–138
76. Nisembaum L, Zigmond M., Sved A, Abercrombie E (1991) Prior exposure to chronic stress results in enhanced synthesis and release of hippocampal norepinephrine in response to a novel stressor. J Neurosci 11: 1478–1484
77. Palfreyman M, Schmidt C, Sorensen S, Dudley M, Kehne J, Moser P, Gittos M, Carr A (1993) Electrophysiological, biochemical and behavioural evidence for 5-HT$_2$ and 5-HT$_3$ mediated control of dopaminergic function. Psychopharmacol 112: S60-S67
78. Pei Q, Zetterström T, Fillenz M (1989) Measurement of extracellular 5-HT and 5-HIAA in hippocampus of freely moving rats using microdialysis: long-term applications. Neurochem Int 15: 503–509
79. Plotsky P, Cunninham E, Widmaier E (1989) Catecholaminergic modulation of corticotropin-releasing factor and adrenocorticotropin szecretion. Endocrine Reviews 10: 437–458
80. Ramage A, Wouters W, Bevan P (1988) Evidence that the novel antihypertensive agent, flesinoxan, causes differential sympathoinhibition and also increases vagal tone by central action. Eur J Pharmacol 151: 373–379
81. Romanowski W, Grabiec S (1974) The role of serotonin in the mechanism of central fatigue. Acta Physiol Pol 25: 127–134
82. Schwartz D, Hernandez L, Hoebel B (1990) Tryptophan increases extracellular serotonin in the lateral hypothalamus of food-deprived rats. Brain Res Bull 25: 803–807
83. Seguin L, Liscia P, Guezennec Y, Fillion G (1998) Effects of moderate and intensive training on functional activity of central 5-HT1B receptors in the rat substantia nigra. Acta Physiologica Scandinavica 162: 63–68
84. Selye H (1936) A syndrome produced by diverse nocuous agents. Nature 138: 32

85. Sharp T, Bramwell S, Grahame-Smith D (1992) Effect of acute administration of L-tryptophan in the release of 5-HT in rat hippocampus in relation to serotonergic neuronal activity: an in vivo microdialysis study. Life Sci 50: 1215–1223

86. Sharp T, Bramwell S, Hjorth S, Grahame-Smith D (1989) Pharmacological characterization of 8-OH-DPAT-induced inhibition of rat hippocampal 5-HT release in vivo as measured by microdialysis. Br J Pharmacol 98: 989–997

87. Shintani F, Nakaki T, Kanba S, Sato K, Yagi G, Shiozawa M, Aiso S, Kato R, Asai M (1995) Involvement of interleukin-1 in immobilisation stress-induced increase in plasma adrenocorticotropic hormones and in release of hypothalamic monoamines in rat. J Neurosci 15: 1961–1970

88. Sleight A, Marsden C, Martin K, Palfreyman M (1988) Relationship between extracellular 5-hydroxytryptamine and behavior following monoamine oxidase inhibition and L-tryptophan. Br J Pharmacol 93: 303–310

89. Stanford C (1993) Monoamines in response and adaptation to stress. In: Stanford S & Salmon P (eds) Stress, from synapse to syndrome. Academic Press, London: 281–331

90. Sutanto W, De Kloet R (1993) The role of GABA in the regulation of the stress response. In: Stanford S & Salmon P (eds) Stress, from synapse to syndrome. Academic Press, London: 333–354

91. Thorré K, Chaouloff F, Sarre S, Meeusen R, Ebinger G, Michotte Y (1997) Differential effects of restraint stress on hippocampal 5-HT metabolism and extracellular levels of 5-HT in streptozotocin-diabetic rats. Brain Research 772: 209–216

92. Tuomisto J, Mannisto P (1985) Neurotransmitter regulation of anterior pituitary hormones. Pharmacology Reviews 37: 249–332

93. Urhausen A, Gabriel H, Kindermann W (1998) Impaired pituitary hormonal response to exhaustive exercise in overtrained endurance athletes. Med Sci Sports Exerc 30: 407–414

94. Ursin H & Ollf M (1993) The stress response. In: Stanford S & Salmon P (eds) Stress, from synapse to syndrome. Academic Press, London: 3–22

95. Weizman R, Weizman A, Kook K, Vocci F, Deitsch S, Paul S (1989) Repeated swim stress alters brain benzodiazepine receptors measured in vivo. J Pharmacol. Exp Ther 249: 701–707

96. Westerink B, De Vries J (1991) Effect of precursor loading on the synthesis rate and release of dopamine and serotonin in the striatum: a microdialysis study in conscious rats. J Neurochem 56: 228–233

97. Wilckens T, Schweiger U, Pirke K (1992) Activation of 5-HT$_{1C}$-receptors suppresses excessive wheel running induced by semi-starvation in the rat. Psychopharmacol 109: 77–84

98. Wilson W, Maughan R (1992) Evidence for a possible role of 5-hydroxytryptamine in the genesis of fatigue in man: administration of paroxetine, a 5-HT re-uptake inhibitor, reduces the capacity to perform prolonged exercise. Exper Physiol 77: 921–924

99. Wurtman R, Lewis M (1991) Exercise, plasma composition and neurotransmission. In: Brouns F (ed) Advances in Nutrition and Top Sport. Med Sport Sci. Basel: Karger 32: 94–109

16

THE ADRENERGIC INFLUENCE ON SLEEP STAGE SHIFTING IN HIGH-ENDURANCE ATHLETES AFTER EXERCISE

Nikolaus C. Netzer, [1]* Hartmut Steinle, [1] Kingman P. Strohl, [2] and Manfred Lehmann[1]

[1.]Department of Sportsmedicine and Rehabilitation
Medical Center
University Hospitals Ulm Steinhövelstr. 9
89071 Ulm, Germany
[2.]Sleep Research Center
Case Western Reserve University
Cleveland, Ohio

1. INTRODUCTION

The sleep of humans and animals is influenced by environmental and inner organic circumstances like the sunlight and circadian rythms of hormones. Sleep by itself can be divided in phases of REM (Rapid Eye Movement)- and Non-REM sleep (1). Both sleep phases influence physical parameters as heartrate, blood pressure, muscle tension and they seem to be both essential for the organism (2). The change of sleep phases is centrally controled in the brain. But at present no single population of neurons and no single transmitter can be made responsible for the control of REM- and Non-REM-sleep. Research of the last three decades however could show that neurons in the gigantic cellular field (FTG) in the pons mainly control REM-Sleep. This part of the brain has mainly cholinoceptive receptors (3, 4, 5, 6) and cholinergic transmitters like carbachol can enforce REM-sleep and anticholinergic substances like atropin can suppress REM-sleep (7, 8). Unlike cholinergic substances the influence of adrenergic and monoaminercig substances in sleep control could not be proven on a neuronal level. However since several years some authors postulate the influence of aminergic substances in sleep control (9, 10). One of their arguments is the strong relation of respiratory neurons and sleep neurons in the same area of the brain and for respiratory neurons the parasympathic as well as the sympathic influence is proven (11, 12).

*Ph.: +49 731 5026966 Fax: +49 731 5056686

Overload, Performance Incompetence, and Regeneration in Sport, edited by Lehmann *et al.*
Kluwer Academic / Plenum Publishers, New York, 1999.

Physical and psychical stress leads to a shift of REM-sleep into a later part of bed- or sleep time and enlarges the so called REM-sleep latency. Several years ago this was postulated as restaurative sleep stage shifting after maximal exercise in athlethes (13). However recent studies and review articles support the idea that a higher level of peripheral adrenergic transmitters like norepinephrine and dopamine—known to cross the blood -brain barrier—in the early part of the night after maximal exercise during daytime lead to a supression of REM-sleep in the first part of the night and a enlargement of the REM-sleep latency (14, 15).

We wanted to show in a trial with high endurance performance athlethes if and how sleep is changed in these persons after maximal exercise in competition compared to physical rest in a non -training phase and if there is any correlation between sleep changes and the peripheral levels of adrenergic transmitters (catecholamines).

2. METHODS

15 male volunteers participated in this study. 8 of them were high performance street cyclists partly of the german national amateur team and 7 trialethes of a german first divison team.

The mean age was 23.9 years (range 19–27y) and the mean body mass index (BMI) was 21.9 kg/meter^2 (range 19.6–25.1). No previous cardial or pulmonary diseases in these athlethes were known. None of the participants had taken any drug including phytotherapeutic drugs in the last six weeks previously to the study and no coffeine or teeine on the day before measurements. All participants were non-smokers. Inclusion criteria was a minimum performance of 400 watts in the cycle exercise test (beginning with 150 watts and steps of 50 watts of 3 minutes each). In the average all participants trained at least 20h per week in their training period. A standard polysomnography was performed in a clinical setting in each participant maximum 5 hours after a competition exercise and another polysomnography during rest in a non training period. The following parameters and channels were measured in the polysomnography (SAC- Sleep System, Oxford Medical Systems): 2 EEG chanels, 2 EOG channels, 2 EMG channels (chin), SaO2 via pulse-oximetry , thoracical and abdominal respiratory effort via belts, ECG, nasal and oral flow via thermistor and body position.

The sleep staging was done manually by one technician using Rechtschaffen and Kales criteria (16). Respiratory events were recorded using the following criteria: An apnea was defined as minimum 80% decrease of airflow for at least 10 seconds (sec), hypopnea was defined as 50% to 80% decrease of airflow for at least 10sec. Obstructive events were defined as decrease or missing airflow without complete missing of abdominal or thoracical respiratory effort, central events were defined as missing airflow and abdominal or thoracical respiratory effort. An oxygen desaturation was defined as decrease of 4% SaO2. Catecholamine levels were measured separately in the daytime (until 10pm)and nighttime urine (until 8am next morning) on the day before and the night during polysomnography. The analysis of the catecholamines in the urine was performed using the analysis method of DaPrada and Zuercher (17). To prove a possible overtraining syndrome the minimum levels for norepinephrine was determined at 10ng/min and for epinephrine at 2–4ng/min. In the overtraining syndrome the levels of catecholamines decrease to 50% of the individual normal level at rest. Statistical evaluation was done using the Student T-test for significance of differences between data after competition and at rest and bivariate correlation is described with the Spearman correlation coefficient for correlations between catecholamine levels and polysomnographic data. A single data is also expressed by descriptive means. Informed consent was obtained from each subject prior to the study and the protocol of this trial was proven by the ethical commitee of the University Freiburg, Germany.

3. RESULTS

3.1. Catecholamine Levels

Two of the 15 athlethes had catecholamine levels on the competition day below the minimum levels because they were in a overtraining syndrome. The following data is expressed without the results of these two athlethes. Daytime catecholamines as expression of catecholamine levels until beginning of the night were significantly different for epinephrine, norepinephrine and dopamine on the competition day (C) and the rest day (R). The average value for epinephrine was 45.5 ng/min (C) versus 17.8 ng/min (R), p< 0.001. For norepinephrine 133.3 ng/min (C) versus 49.2 ng/min (R), p= 0.004, and for dopamine 609.9ng/min (C) versus 223.7 ng/min (R), p= 0.03. Nighttime levels were significantly different for epinephrine and norepinephrine but not for dopamine. Epinephrine was 15.7ng/min (C) versus 8.4 ng/min (R), p= 0.006; norepinephrine 54.1ng/min (C) versus 28.4ng/min (R), p= 0.03; dopamine 304.6ng/min (C) versus 221.9 (R), p= 0.14.

3.2. Sleep Staging

In the comparison of polysomnographic data between competition day and rest day the total sleep stage values as percentage of total sleep time (TST) were significantly different only for stage 3 Non-Rem- sleep: In the average stage 3 was 13.9% of TST (C) versus 11.4% of TST (R), p= 0.03. REM showed a tendency to be decreased on competition day: 15.5% of TST (C) versus 17.6% of TST (R), p = 0.09. In the first two Non-REM/REM cycle of the night however REM was significantly decreased (p=0.05) and stage 2 Non-REM sleep significantly increased (p=0.03) on competition day versus rest day, REM then showed a tendency to be decreased in the second two Non-REM/REM cycles of the night (p= 0.1). The absolute sleep latency (wake until stage 1) was not different between competition day and rest day [19.1 min (C) versus 19.2 min (R)]. However the REM onset latency was significantly different [124.0 min (C) versus 97.2min (R), p = 0.05]. There were no differences at all for sleep efficiency (total sleep time /sleep period - or bed- time).

3.3. Respiratory Analysis

No differences for respiratory data were found between competition and rest day. Except one individual the athlethes had no respiratory events and oxygen desaturations at all or less than 10 events and desaturations during the whole night (TST).

3.4. Heartfrequency Analysis

The heartfrequency was significantly different at the start of bedtime between competition and rest day: 54.7 /min (C) versus 50. 5/min (R). During the rest of the night the heartfrequency was not significantly different between competition and rest day.

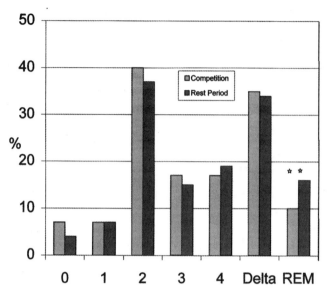

Figure 1. Sleep stages in percent of total sleep time in the first half of the night (first two REM/Non-REM cycles) Black bars: After competition excersise, White bars: In training and competition free period (** = p < 0.05).

3.5. Correlations between Catecholamine Levels and Sleep Staging Data

The only significant correlations between sleep staging and catecholamine levels were found for REM- sleep onset latency and catecholamine levels and for REM distribution in the first two REM/Non-REM cycles of the night and catecholamine levels.

The spearman correlation coefficient was r = 0.53 (p= 0.04) for the changes of REM- sleep latency and epinephrine between competition to rest daytime urine level. For

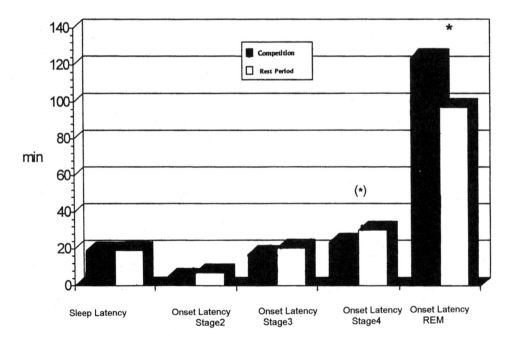

Figure 2. Onset latencies of sleep stages in minutes (* p= 0.05).

norepinephrine nighttime urine levels and REM- onset latency the correlation coefficient was r = 0.45 (p= 0.09). Absolutely seen without comparing differences between competition day and rest day there was a correlation between norepinephrine levels in nighttime urine and REM-onset latency (r = 0.36, p= 0.06), there was only a mild correlation for epinephrine in nighttime urine and REM-onset latency (r = 0.26, p= 0.19). The correlation between REM percentage in the first two REM/Non-REM cycles and nightime urine levels for norepinephrine was r = - 0.46 (p < 0.01), for daytime norepinephrine r = -0.46 (p< 0.01), for nighttime epinephrine r = -0.35 (p= 0.06) and for daytime epinephrine r = -0.35 (p = 0.06).

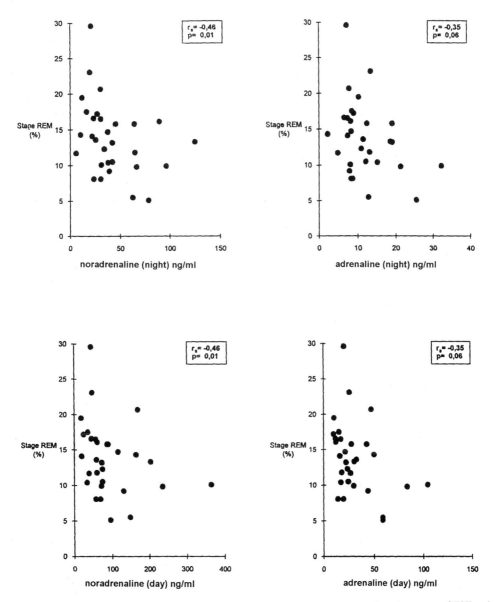

Figure 3. Correlations between daytime and nighttime epinephrine levels and REM sleep in percent of TST and daytime and nighttime norepinephrine levels and REM-sleep (% TST) independent of day of polysomnography.

No correlations were found between sleep stage percentages of total sleep time or their changes and catecholamine levels.

4. DISCUSSION

The results of this study showed a moderate change of sleep stages in the polysomnography of high performance athlethes on a day after maximal competition exercise compared to a day in a non training period without exercise. The most significant changes were found for REM-sleep onset latency and the distribution of REM during the night. A shift of REM-sleep from the first to the second half of the night occurs. These changes of REM- sleep distribution during night after the competition day are correlated with increased peripheral catecholamine levels of epinephrine and norepinephrine and a increased heart rate on the day and during the night after competition exercise. These results support earlier findings in other studies, which reported REM- sleep shift and decreased REM after exercise in athlethes and normal persons (18, 19). However it does not seem as if major changes in sleep or an improved or worsened quality of sleep would be triggered by maximal exercise. The major quality of sleep defined by sleep efficiency and the total distribution of sleep stages did not change in our subjects after maximal exercise versus no exercise in a non training period. These findings are opposite to reports of previous studies with improved sleep quality after exercise and restaurative sleep (13, 20). It must be critizised however that many of these findings in previous studies where not based on full polysomnography but on self reported sleep quality. Studies with polysomnographic measurements come to comparable results as in our trial.

We found partly significant correlations between REM-sleep shift and peripheral catecholamine levels in daytime and nighttime urine. These findings are limited by the fact, that peripheral adrenergic transmitter levels do not necessarely represent central respectively cerebral levels of these transmitters. However there are studies that report strong correlations between peripheral and central levels of these transmitter as some of these catecholmines, especially norepinephrine, can cross the blood-brain barrier. The increased cathecholamine levels in our subjects also had a strong systemic impact showing a significantly increased heartfrequency. The correlations between REM-sleep shift and increased catecholamine levels support the hypothesis that REM sleep generation in the brain is not only triggered by cholinergic transmitters but may also be supressed by the influence of aminergic transmitters (21). Therefore physical stress like maximal exercise may lead to a REM-sleep supression in the first part of the night via increased catecholamine levels. However this does not decrease sleep quality as there seems to be a REM-rebound in the second half of the nigth when catecholamine levels decrease and cholinergic transmitters relatively increase. This hypothesis supported by our findings and other studies in open trials with humans must however be proven by research on a cerebral level in animal basic research before further conclusions can be drawn.

REFERENCES

1. Carscadon MA, Dement WC. Normal Human Sleep (1994) An Overview. In: MH Kryger, T Roth, WC Dement (Eds.). Principles and Practices of Sleep Medicine. 2nd Ed.WB Saunders Philadelphia 16–25
2. Rechtschaffen A, Bergman BM, Everson CA, Kushida CA, Gilliland MA (1989) Sleep deprivation in the rat: integration and discussion of the findings. Sleep 12: 68–87

3. Jouvet M (1972) The role of monoamines and acethylcholine-containing neurons in the regulations of the sleep waking cycle. Ergebn Physiol 64: 166–307
4. Hobson JA (1974) The cellular basis of sleep cycle control. In Advances in Sleep Research, Vol 1. Ed. ED Weitzman. Spectrum New York 217–249
5. Hobson JA, Steriade M (1986) Neuronal basis of behavioral state control. In Handbook of physiology. The Nervous System. Intrinsic Regulatory Systems of the Brain. Sect. 1, Vol. 4 Am. Physiol. Soc. Bethesda 701–823
6. Hobson JA, McCarley RW, Freedman R, Pivik RT (1974) Time course of discharge rate changes by cat pontine brainstem neurons during the sleep cycle. J Neurophysiol. 37: 1297–1309
7. Amatruda TT, Black DA, McKenna TM, Carley RW, Hobson JA (1975) Sleep cycle control and cholinergic mechanisms: different effects of carbachol injections at pontine brain stem sites. Brain Res. 98:501–515
8. Baghdoyan HA, Lydic R, Callaway CW, Hobson JA (1989) The Carbachol-induced enhancement of desynchronized sleep signs is dose dependent and antagonized by centrally administered atropine. Neuropsychopharmacology 2: 67–69
9. McCarley RW (1980) Mechanisms and models of behavioral state control. In: JA Hobson and MAB Brazier (Eds.).The Reticular Formation Revisited. Raven Press New York 375–403
10. Lydic R, McCarley RW, Hobson JA (1984) Forced activity alters sleep cycle periodicity and dorsal raphe discharge rythm. Am J. Physiol. 247: 135–145
11. Steriade M, McCarley RW (1990) Brainstem Control of Wakefulness and Sleep. Plenum Press New York
12. Burton MD, Johnson DC, Kazemi H (1990) Adrenergic and cholinergic interaction in ventilatory control. J. Appl. Physiol 68:2092- 2099
13. Shapiro CM, Bortz R, Mitchell D, Bartel P, Jooste P (1981) Slow wave sleep : a recovery period after exercise. Science 214: 1253–1254
14. Mistlberger R, Bergmann B, Rechtschaffen A (1987) Period -amplitude analysis of rat electroencephalogram: effects of sleep deprivation and exercise. Sleep 10: 508–522
15. Torsvall L, Akerstedt T, Lindbeck G (1984) Effects on sleep stages and EEG power density of different degrees of exercise in fit subjects.Electroencephalogr Clin Neurophysiol 57: 347–353
16. Rechtschaffen A, Kales A (1968) A manual of standardized terminology, techniques and scoring system for sleep stages in human subjects. US Departement of Health, National Institute of Neurological Disease and Blindness, Bethesda/MD
17. DaPrada M, Zürcher G (1976) Simultaneous radioenzymatic determination of plasma and tissue adrenalin, noradrenalin an dopamine within the fentomole range. Life Sci 19: 1161–1174
18. Matsumoto K, Saito Y, Abe M, Furumi K (1984) The effects of daytime exercise on night sleep. J Hum Ergol 13: 31–36
19. Taylor SR, Rogers GG, Driver HS (1997) Effects of training volume on sleep, psychological, and selected physiological profiles of elite female swimmers. Med Sci Sports Exerc 29:688–693
20. King AC, Oman RF, Brassington GS, Bliwise DL, Haskell WL (1997) Moderate- intensity exercise and self -rated quality of sleep in older adults. A randomized controlled trial. J Am Med Assoc 277:32–37
21. Lydic R, Baghdoyan HA (1994) The Neurobiology of Rapid Eye Movement Sleep. In: Sullivan CE, Saunders NA (Eds.).Sleep and Breathing. Marcel Dekker New York 47- 77

INDEX